T0305864

Forms of Fermat Equations
and Their Zeta Functions

Forms of Fermat Equations and Their Zeta Functions

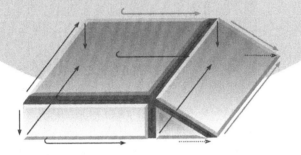

Lars Brünjes
University of Regensburg, Germany

W⊖ World Scientific

NEW JERSEY • LONDON • SINGAPORE • BEIJING • SHANGHAI • HONG KONG • TAIPEI • CHENNAI

Published by

World Scientific Publishing Co. Pte. Ltd.

5 Toh Tuck Link, Singapore 596224

USA office: 27 Warren Street, Suite 401–402, Hackensack, NJ 07601

UK office: 57 Shelton Street, Covent Garden, London WC2H 9HE

British Library Cataloguing-in-Publication Data
A catalogue record for this book is available from the British Library.

FORMS OF FERMAT EQUATIONS AND THEIR ZETA FUNCTIONS

ISBN-13 978-981-256-039-1
ISBN-10 981-256-039-4

Printed in Singapore

Dedicated to my wife

June Roberts

whose patience and support made this book possible

Preface

This book is a translated and modified (hopefully improved) version of the author's PhD-thesis "Über die Zetafunktion von Formen von Fermatglei- chungen". Apart from correcting minor mistakes, explanations and exam- ples as well as chapters recalling basic definitions on zeta functions and l-adic cohomology have been added to assist readers that are not experts in the field.

I want to express my gratitude towards the people without whose help this book could not have been written: Prof. Dr. Uwe Jannsen, the su- pervisor of my PhD-thesis, who always showed a keen interest in my progress and was always willing to give advise, my best friend and col- league Dr. Kirsten Schneider, who sacrificed many hours to listen to my problems and to give valuable hints, my colleague Dr. Christopher Rup- precht, to whom I owe most of what I know about the classification of Fermat equations, the staff of World Scientific Publishing, who was very helpful and patient, and of course my wife June Roberts, whose trust, love and patience have been essential for me during the last years.

Lars Brünjes, Regensburg 2004

Contents

Chapter 1

Introduction

One way to describe a mathematical discipline is to describe the equations studied in that discipline. General algebraic geometry is concerned with polynomial equations over arbitrary (commutative) rings, whereas number theory and arithmetic geometry specialize in polynomial equations over number fields, finite fields, p-adic fields and rings of integers.

A particularly famous example is the *Fermat equation* P_n^r over a ring R

$$X_1^r + X_2^r + \ldots + X_n^r = 0$$

for natural numbers $r, n \geq 2$ and variables $X_1, \ldots, X_n \in R$. The special case $r \geq 3$, $n = 3$, $R = \mathbb{Z}$ is subject of *Fermat's Last Theorem*, for hundreds of years the most famous mathematical conjecture, proved by Andrew Wiles in 1993.

In his way-breaking 1949 paper [Wei49], André Weil studied the special case where R is a finite field $\mathbb{F}_q$, using this to motivate his *Weil conjectures*, whose proof, completed by Deligne in 1974 ([Del74]), was undoubtedly one of the great triumphes of 20th century mathematics.

In modern algebraic geometry, Fermat equations (respectively *Fermat hypersurfaces*, the hypersurfaces defined by Fermat equations in projective space) have been studied intensely, especially by Tetsuji Shioda ([SK79], [Shi79], [Shi82], [Shi83], [Shi87], [Shi88]), and they are often used as examples and testing ground for open problems like the conjectures of Hodge and Tate.

But even though much more is known about Fermat hypersurfaces than about general hypersurfaces, not to mention general varieties, a lot of questions still remain open; both the Hodge and the Tate conjecture for example, though proven by Shioda for a large class of Fermat hypersurfaces, still remain open for general Fermat hypersurfaces.

Now let $R = k = \mathbb{F}_q$ be a finite field. Then every homogenous equation $f(X_1, \ldots, X_n) = 0$ over k has only a finite number $\nu^{(i)}$ of solutions in the finite extensions $\mathbb{F}_{q^i}$ of k, and the collection of all these numbers in the *zeta function*

$$\zeta(f, T) := \exp\left(\sum_{i=1}^{\infty} \frac{\nu^{(i)}}{i} T^i\right) \in \mathbb{Q}[[T]],$$

is one of the most fundamental and important invariants of f respectively the $(n-2)$-dimensional hypersurface $X(f)$ defined by f in $\mathbb{P}_k^{n-1}$.

A finer invariant is the l-adic cohomology $\mathrm{H}^*_{\text{ét}}(\bar{X}(f), \mathbb{Q}_l)$ of X (for $\bar{X}(f) := X(f) \times_k \bar{k}$ and a prime $l \neq \mathrm{char}\,(k)$), a finite dimensional $\mathbb{Q}_l$-linear representation of the absolute Galois group G_k. Knowing this representation means knowing the zeta function because of the formula

$$\nu^{(i)} = \sum_{j=0}^{2(n-2)} \mathrm{Tr}\left[(F^*)^i \Big| \mathrm{H}^j_{\text{ét}}(\bar{X}(f), \mathbb{Q}_l)\right],$$

where $F \in G_k$ is the *geometric Frobenius* in G_k, the inverse of the k-automorphism $x \mapsto x^q$.

The zeta function of the Fermat hypersurface $X_n^r := X(P_n^r)$ was already known to Weil; he computed it in the above mentioned article and showed in particular that it was not only a power series but a *rational* function, leading him to the conjecture that the zeta function of smooth, projective varieties over k is always rational.

Later Deligne computed the Galois-representation $\mathrm{H}^*_{\text{ét}}(\bar{X}_n^r, \mathbb{Q}_l)$ (see [Del82]), which is particularly simple, since the cohomology decomposes into canonical one-dimensional subspaces on which the Frobenius acts by certain Hecke characters, the so called *Jacobi sums*.

Taking Fermat hypersurfaces as a starting point, it is a natural step to consider slightly more general hypersurfaces like *diagonal hypersurfaces*, given by *diagonal* equations of the form

$$a_1 X_1^r + \ldots + a_n X_n^r = 0$$

for $a_i \in k$. These were studied thoroughly (for $k = \mathbb{F}_q$, $q \equiv 1 \pmod{r}$) by Fernando Q. Gouvêa und Noriko Yui in the book [GY95]; using Weil's and Deligne's results, it is not difficult to compute the zeta function of such diagonal equations.

"Geometrically", i.e. over $\bar{k}$, every diagonal equation of degree r in n unknowns is isomorphic to the Fermat equation P_n^r, in the sense that there is a linear change of variables $(X_i \mapsto \sqrt[r]{a_i}X_i)$ transforming the given equation into P_n^r. Thus diagonal equations are only interesting if k is not algebraically closed, so that "arithmetical" questions come into play.

Look at the following two quadratic diagonal equations in three unknowns over $\mathbb{Q}$:

$$P_3^2 : X_1^2 + X_2^2 + X_3^2 = 0 \quad \text{and}$$
$$Q : \ X_1^2 - 2X_2^2 - X_3^2 = 0.$$

Whereas P_3^2 has no non-trivial solution in $\mathbb{Q}$, equation Q has infinitely many ("Pell's equation"), so even though P_3^2 and Q become isomorphic over $\bar{\mathbb{Q}}$, they seem to be completely unrelated over $\mathbb{Q}$.

But in fact it *is* possible to use the isomorphism over $\bar{\mathbb{Q}}$ to gain information over $\mathbb{Q}$ by a general principle known as *Galois descent*: If K/k is a Galois extension with Galois group G, and if X is an object "defined over k", then every object Y over k which becomes isomorphic to X "over K" (and is then called a *K/k-form of X*) defines a class in $H^1(G, A(X))$, where $A(X)$ is the (not necessarily abelian) automorphism group of X over K. The idea of descent is to deduce properties of Y from properties of X by "twisting" with this class. In particular this can be done in the case of diagonal equations, since they are $\bar{k}/k$-forms of Fermat equations.

This immediately leads to the following question: *Are there $\bar{k}/k$-forms of Fermat equations, so-called "twisted Fermat equations", which are not diagonal?* — If the answer is "yes", one can then try to use Galois descent to answer questions about the cohomology and zeta function of such twisted equations.

In the case $K = \bar{k}$, the automorphism group $A(P_n^r)$ equals $\mu_r \wr S_n$, the wreath product of the group of r-th roots of unity in K with the symmetric group S_n; here S_n acts on P_n^r by permuting the X_i, and $(\zeta_i) \in \mu_r^n$ acts by $X_i \mapsto \zeta_i X_i$.

The K/k-forms of P_n^r are then given by classes in $H^1(G_k, \mu_r \wr S_n)$, and it turns out that the diagonal equations are precisely those whose class already comes from $H^1(G_k, \mu_r^n)$, so that from this point of view, considering diagonal equations alone is an unnatural restriction.

In this book we therefore want to equally consider *all* forms of P_n^r, first classify them, then study them with the method of descent and (in the case $k = \mathbb{F}_q$) compute their zeta function.

In contrast to the case of diagonal equations, it is difficult to see in general whether a given equation is a form of the Fermat equation or not — the equation

$$4X_1^2 X_2 + 3X_1 X_2^2 + 3X_1^2 X_3 + 4X_2^2 X_4 + 4X_1 X_3 X_4 + X_1 X_4^2$$
$$+ 4X_2 X_3^2 + X_2 X_3 X_4 + X_3 X_4^2 + 2X_3^3 + X_3^2 X_4 + 2X_3 X_4^2 + 3X_4^3 = 0$$

for example is a form of the Fermat equation P_4^3 over the field $\mathbb{F}_5$ which is *not* diagonal. And even for equations as complicated as this one the method of descent will enable us to compute the group of automorphisms over k and the zeta function.

On the other hand, it is of course a slight disadvantage of general twisted Fermat equations that they do not show any apparent symmetry, because this fact often makes it difficult to decide whether a given equation is a twisted Fermat equation or not.

An interesting exception is the case $r = 3$, $n = 2$, the case of *binary cubic equations*, because in that case *all* "non-singular" equations are forms of P_2^3. Using the methods explained in this book, we can not only classify all such forms, but make this classification completely explicit, enabling us to compute the class and (if k is a finite field) the zeta function for any given (non-singular) binary cubic equation.

In the *second chapter* we will introduce one of the key concepts of this book, the *zeta function* of a variety over a finite field, and we will state the Weil-conjectures.

Even though the method of descent is "folklore", the details can be somewhat tricky. Therefore in the *third chapter*, we explain what we mean by "Galois descent" and axiomatize an important class of situations in which Galois descent holds:

Definition: A *coefficient extension (from k to K)* consists of two categories $\mathcal{C}_k$ and $\mathcal{C}_K$, a covariant functor $F : \mathcal{C}_k \to \mathcal{C}_K$ and for all $Y, Z \in \mathrm{Ob}(\mathcal{C}_k)$ a G-action (from the left) on $\mathrm{Iso}_{\mathcal{C}_K}(FY, FZ)$ (the set of isomorphisms between FY and FZ in $\mathcal{C}_K$), so that the following two conditions are satisfied:

(CE1): The action is compatible with compositions, i.e. for objects $X, Y, Z \in \mathrm{Ob}(\mathcal{C}_k)$, isomorphisms $X \xrightarrow{g} Y$ and $Y \xrightarrow{f} Z$ and an element $s \in G$, we have:

$$^{s}(fg) = {}^{s}f {}^{s}g.$$

(CE2): Exactly those isomorphisms that come from $\mathcal{C}_k$ are fix under the action of G, i.e. for objects $Y, Z \in \mathrm{Ob}(\mathcal{C}_k)$ we have:

$$\mathrm{Im}\left(\mathrm{Iso}_{\mathcal{C}_k}(Y, Z) \xrightarrow{F} \mathrm{Iso}_{\mathcal{C}_K}(FY, FZ)\right) = \left[\mathrm{Iso}_{\mathcal{C}_K}(FY, FZ)\right]^{G}.$$

Two examples of coefficient extensions are particularly important for our study of K/k-forms of Fermat equations: First the categories $\mathcal{F}_k^{n,r}$ and $\mathcal{F}_K^{n,r}$ whose objects are homogenous equations of degree r in n unknowns with coefficients in k respectively K and whose morphisms are elements of $\mathrm{GL}(n, k)$ respectively $\mathrm{GL}(n, K)$, considered as linear changes of variables. The coefficient extension is then given by the obvious functor $\mathcal{F}_k^{n,r} \to \mathcal{F}_K^{n,r}$ and the natural G-action on $\mathrm{GL}(n, K)$.

The importance of this coefficient extension for us is obvious: The Fermat equation P_n^r is an object of $\mathcal{F}_k^{n,r}$, and the twisted Fermat equations are exactly those objects of $\mathcal{F}_k^{n,r}$ that become isomorphic to P_n^r in $\mathcal{F}_K^{n,r}$.

The second important example is given by the categories $\mathbf{Rep}_{\mathbb{Q}_l}^{G_k}$ and $\mathbf{Rep}_{\mathbb{Q}_l}^{G_K}$ of $\mathbb{Q}_l$-G_k-representations respectively $\mathbb{Q}_l$-G_K-representations, together with the functor $\mathbf{Rep}_{\mathbb{Q}_l}^{G_k} \to \mathbf{Rep}_{\mathbb{Q}_l}^{G_K}$ that maps a representation $G_k \xrightarrow{\varphi} \mathrm{Aut}_{\mathbb{Q}_l}(V)$ to its restriction $\varphi|G_K$. Here the action of an element $\bar{s}$ of $G = G_k/G_K$ on $(V, \varphi) \xrightarrow{f} (W, \psi)$ is defined by "conjugation", i.e. by $f \mapsto \psi(s) f \varphi(s)^{-1}$.

This coefficient extension is important for us, because the l-adic cohomology $\mathrm{H}_{\text{ét}}^{*}(\bar{X}_n^r, \mathbb{Q}_l)$ is an object of $\mathbf{Rep}_{\mathbb{Q}_l}^{G_k}$ and because the cohomology of a form of the Fermat hypersurface X_n^r is a form of $\mathrm{H}_{\text{ét}}^{*}(\bar{X}_n^r, \mathbb{Q}_l)$.

Although these two examples are the most important for our purposes, coefficient extensions occur in many more situations, and we will give some interesting additional examples like the base change functor from the category of k-varieties to the category of K-varieties.

As already mentioned above, the classification of forms using Galois descent will involve the cohomology $H^1(G, A)$ for not necessarily abelian coefficients A, because the automorphism groups of objects we study will often be nonabelian — the wreath product $A(P_n^r) = \mu_r \wr S_n$ for example is not abelian for $n \geq 2$. Therefore in the *fourth chapter* we want to explain the basic definitions and results from the theory of nonabelian cohomology. We will closely follow Serre's presentation from [Ser97], being slightly more general by considering general topological groups G and not only profinite ones. This will in particular enable us to treat our Galois group G as both a discrete and a profinite group.

The *fifth chapter* will give a brief introduction to l-adic cohomology and Weil cohomology theories and explain how these topics are related to the Weil conjectures and their proof. In particular, we will see how the zeta function of a variety X can be computed from the Galois-action on the l-adic cohomology $\mathrm{H}_{\text{ét}}^*(\bar{X}, \mathbb{Q}_l)$ of X.

In the *sixth chapter* we are going to explain how to use Galois descent in the framework of arbitrary coefficient extensions $F : \mathcal{C}_k \to \mathcal{C}_K$ to classify forms: If X is an object of $\mathcal{C}_k$, then by $E(\mathcal{C}_K/\mathcal{C}_k, X)$ we denote the set[2] of $\mathcal{C}_K/\mathcal{C}_k$-*forms of* X, i.e. of isomorphism classes $[Y]$ of objects of $\mathcal{C}_k$ such that FY is isomorphic to FX in $\mathcal{C}_K$.

If $[Y]$ is such a $\mathcal{C}_K/\mathcal{C}_k$-form of X, and if $FY \xrightarrow{f} FX$ is an isomorphism in $\mathcal{C}_K$, then $s \mapsto f^s(f^{-1})$ defines a 1-cocycle $a = (a_s)$ of $G := \mathrm{Gal}(K/k)$ in $A(X) := \mathrm{Aut}_{\mathcal{C}_K}(FX)$ whose cohomology class $\vartheta[Y]$ in $H^1(G, A(X))$ will turn out to be independent of Y and f:

> **Proposition:** *The assignment* $[Y] \mapsto \vartheta[Y]$ *is a well defined injection from* $E(\mathcal{C}_K/\mathcal{C}_k, X)$ *into* $H^1(G, A(X))$.

As a first application we will compute the group of automorphisms of Y in $\mathcal{C}_k$: It is $\mathrm{Aut}_{\mathcal{C}_k}(Y) = (A(X)_{\vartheta[Y]})^G$, where $A(X)_{\vartheta[Y]}$ is the group $A(X)$

---

[2]A priori this is only a class, but we will prove that it is indeed a set

with its G-action *twisted by the class* $\vartheta[Y]$, i.e. $s \in G$ acts by $b \mapsto a_s {}^s b a_s^{-1}$.

If $\mathcal{C}_k' \to \mathcal{C}_K'$ is another coefficient extension, and if we have functors $\mathcal{C}_k \xrightarrow{H_k} \mathcal{C}_k'$ and $\mathcal{C}_K \xrightarrow{H_K} \mathcal{C}_K'$ which are compatible with each other and with the G-actions (a notion to be made precise, of course), we talk about a *morphism of coefficient extensions*, and we get a commutative diagram

$$
\begin{array}{ccc}
E(K/k, X) & \xrightarrow{\ [Y] \mapsto [H_k Y]\ } & E(K/k, H_k X) \\[2mm]
\Big\downarrow{\vartheta} & = & \Big\downarrow{\vartheta} \\[2mm]
H^1(G, A(X)) & \xrightarrow{\ (a_s) \mapsto (H_K a_s)\ } & H^1(G, A(H_k X))
\end{array}
$$

This result is very important for our purposes if we apply it to the coefficient extensions $\mathcal{F}_k^{n,r} \to \mathcal{F}_K^{n,r}$ and $\mathbf{Rep}_{\mathbb{Q}_l}^{G_k} \to \mathbf{Rep}_{\mathbb{Q}_l}^{G_K}$ introduced above and the functors H_k and H_K induced by l-adic cohomology, because it then states the following:

If Q is a $\mathcal{F}_K^{n,r}/\mathcal{F}_k^{n,r}$-form of the Fermat equation P_n^r, characterized by a cohomology class $\vartheta[Q]$, then the cohomology of the hypersurface $X(Y)$ and hence its zeta function are determined by the class $H_{\text{ét}}^* \vartheta[Q]$.

So instead of having to compute the cohomology of $X(Q)$ directly, to get the zeta function, we can compute the composition $\vartheta^{-1} \circ H_{\text{ét}}^* \circ \vartheta$ which explicitly involves the following steps:

- We have to understand the l-adic cohomology of the Fermat hypersurface X_n^r with its Galois action — this is achieved by Deligne's result mentioned above.
- For an automorphism $a \in A(P_n^r)$, we have to compute the associated automorphism $H_{\text{ét}}^*(a)$ on $H_{\text{ét}}^*(\bar{X}_n^r, \mathbb{Q}_l)$.
- We must compute the preimage of any given class under the injection ϑ.

The *seventh chapter* is devoted to the study of the nonabelian cohomology $H^1(G, \mu_r \wr S_n)$ and hence to the study of $\mathcal{F}_K^{n,r}/\mathcal{F}_k^{n,r}$-forms of P_n^r for $K := \bar{k}$. We mainly follow the lines of Christopher Rupprecht's diploma thesis [Rup96], making the maps and constructions involved more explicit though in order to later allow us to compute zeta functions explicitly.

Furthermore, we are going to compute the groups of automorphisms of twisted Fermat equations.

The main results from that chapter are:

If $r \geq 3$, we have a bijection

$$E(\mathcal{F}_K^{n,r}/\mathcal{F}_k^{n,r}, P_n^r) \cong \coprod_L \mathrm{Aut}_k(L)\backslash(L^\times/L^{\times r})$$

where the disjoint union is taken over k-isomorphism classes of separable k-algebras L of degree n over k. If Q denotes the equation corresponding to the pair (L, x) under this bijection and if $L = \prod_{i=1}^m L_i$ with fields L_i/k, then we have the following canonical exact sequence of groups:

$$1 \to \prod_{i=1}^m (L_i \cap \mu_r) \to \mathrm{Aut}_{\mathcal{F}_k^{n,r}}(Q) \to$$

$$\left\{ a \in \mathrm{Aut}_k(L)^{\mathrm{opp}} \,\middle|\, \frac{ax}{x} \in L^{\times r} \right\} \to 1.$$

In *chapter eight* we will look at the special case of binary cubic equations of a field k with $\mathrm{char}\,(k) \geq 5$. As already mentioned above, in that case *all* non-singular objects of $\mathcal{F}_k^{2,3}$ (i.e. those with non vanishing discriminant) are forms of the Fermat equation P_2^3.

As an application of chapter seven, we are first going to give a complete list of $\mathcal{F}_K^{2,3}/\mathcal{F}_k^{2,3}$-forms of P_2^3 for *finite* fields k. Then we will show how the classification achieved in chapter seven can be made totally explicit in $\mathcal{F}_k^{2,3}$: For a given non-singular binary cubic equation, the corresponding pair (L, x) from the bijection above can be computed as follows:

> **Theorem:** *Let $Q(X,Y) = aX^3 + bX^2Y + cXY^2 + dY^3$ be a non-singular equation over k with $a \neq 0$. Put*
>
> $$\delta := -\frac{\Delta(Q)}{27} \qquad\qquad \in k^\times,$$
>
> $$e := \frac{a}{2} - \frac{27a^2 d + 2b^3 - 9abc}{2\Delta(Q)}\sqrt{\delta} \in k(\sqrt{\delta})^\times,$$
>
> *where*
>
> $$\Delta(Q) := -27a^2 d^2 + 18abcd + b^2 c^2 - 4b^3 d - 4ac^3$$
>
> *is the* discriminant *of Q and where the square root of δ has to be chosen in such a way that $e \neq 0$.*

Then Q corresponds to the pair:

$$\begin{cases} \left(k \times k, \left(e, \dfrac{\sqrt{\delta}}{e}\right)\right) & \text{if } \delta \in k^{\times 2}, \\[2ex] (k(\sqrt{\delta}), e) & \text{otherwise.} \end{cases}$$

The real polynomials $x^4 + y^4 + z^4$ and $-x^4 - y^4 - z^4$ are not isomorphic in $\mathcal{F}_{\mathbb{R}}^{3,4}$, i.e. there is no linear change of variables with coefficients in $\mathbb{R}$ that transforms the one into the other.

If we interpret polynomials as equations, then this fact seems unnatural, since both polynomials obviously have the same set of solutions and "ought to be" isomorphic. Following this intuition, in *chapter nine* we are going to consider a slightly modified coefficient extension $\widetilde{\mathcal{F}_k^{n,r}} \to \widetilde{\mathcal{F}_K^{n,r}}$ in which polynomials that only differ by a scalar become isomorphic. Morphisms in these categories are no longer elements of $\mathrm{GL}\,(n, k)$ respectively $\mathrm{GL}\,(n, K)$ but of $\mathrm{PGL}\,(n, k)$ respectively $\mathrm{PGL}\,(n, K)$. It seems plausible that there will be "less" $\widetilde{\mathcal{F}_K^{n,r}}/\widetilde{\mathcal{F}_k^{n,r}}$-forms of P_n^r than $\mathcal{F}_K^{n,r}/\mathcal{F}_k^{n,r}$-forms, and we are going to show how to make this intuitive notion precise:

> **Proposition:** *Two pairs (L, x) and (L, x') define the same form in $\widetilde{\mathcal{F}_k^{n,r}}$ if and only if there is $a \in Aut_k(L)$, $\lambda \in k^\times$ and $y \in L^\times$, such that $x' = a[\lambda x y^r]$.*

The fact that we have an action of the wreath product $\mu_r \wr S_n$ on X_n^r implies that there is a decomposition of the l-adic cohomology of X_n^r into eigenspaces corresponding to the characters of the abelian group μ_r^n.

In the *tenth chapter* we will more generally study the situation of a semidirect product $A \rtimes S$ (A being finite abelian) acting on an object M of a pseudo-abelian category. It will turn out that again we get a decomposition of M into eigenspaces M_χ corresponding to the characters χ of A, and if p_χ denotes the injection $M_\chi \to M$, then for any $s \in S$ we get an induced morphism $s_\chi : M_{s_\chi} \to M_\chi$ such that the following diagram commutes:

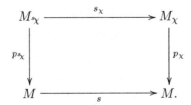

So we see that the decomposition of $H^*_{\text{ét}}(\bar{X}^r_n, \mathbb{Q}_l)$ into eigenspaces is a "motivic" one, the l-adic realization of a corresponding decomposition of the Grothendieck motif $h(X^r_n)$ of X^r_n.

Chapter eleven is the technical heart of this book where we will compute the isomorphism induced on the l-adic cohomology of X^r_n by an element $\tau \in \mu_r \wr S_n$.

The result for $\tau \in \mu^n_r$ is already known from chapter ten, so that we can concentrate on $\tau \in S_n$ and even reduce further to the case where τ is the transposition (12).

As a result we will be able to prove the existence of a basis $\{v_a\}_a$ of $H^{n-2}_{\text{ét}}(\bar{X}^r_n, \mathbb{Q}_l)$ in terms of which we explicitly know the action of $\mu^n_r \wr S_n$.

In the papers of Shioda, Gouvêa und Yui, it is always assumed that the base field k contains the r-th roots of unity, and in this case the Galois action with respect to the basis $\{v_a\}_a$ is well known, so that we can use the results from chapter six to explicitly describe the Galois action on the l-adic cohomology of any $\mathcal{F}^{n,r}_K/\mathcal{F}^{n,r}_k$-form of P^r_n.

In addition to that, we will be able to show that it is often possible to do the same in the general case over arbitrary finite base fields.

In *chapter twelve* we will sum up the results needed for the computation of the zeta function of a given twisted Fermat equation, and we will give a step-by-step explanation of how to use these results for explicit computations, finally illustrating everything with an instructive example.

At the end of the book, two appendices can be found, one explaining some basic definitions and facts about *pseudo-abelian categories*, the other explicitly computing a certain *Jacobi sum* needed in the examples.

Chapter 2

The zeta function

In this short chapter, we want to define the *zeta function* of a variety over a finite field and quote the *Weil conjectures*.

Definition 2.1   Let k be a field. A *k-variety* or *variety over* k is a separated, reduced k-scheme of finite type (not necessarily irreducible). We denote the category of k-varieties (whose morphisms are morphisms of k-schemes) by $\mathbf{Var}_k$.

Readers not familiar with the notion of schemes can consult one of the standard textbooks on algebraic geometry like [Har93] or [Mum94]. Everything can of course also be found in Grothendieck's *Éléments de géométrie algébrique (EGA)*, the "Bible" of algebraic geometry.

For the rest of this chapter, let $k = \mathbb{F}_q$ be a finite field of characteristic $p > 0$.

Let X be a k-variety, and let k'/k be a finite extension. By definition, a k'-rational point of X is a k-morphism $\operatorname{Spec} k' \to X$, and the set of all k'-rational points of X is denoted by $X(k')$.

Remember that if $X = \operatorname{Spec}\left[k[X_1,\ldots,X_n]/(f_1,\ldots,f_m)\right]$ is *affine*, then a k'-rational point is given by a simultaneous solution of the equations f_j in k', i.e. by a tuple $(x_1,\ldots,x_n) \in k'^n$ with $f_j(x_1,\ldots,x_n) = 0$ for all j. We call this tuple the *affine coordinates* of the corresponding k'-rational point.
Remember further that if $X \subseteq \mathbb{P}_k^n$ is *projective*, given by homogeneous equations $f_1,\ldots,f_m \in k[X_0,\ldots,X_n]$, which means that $X =$

Proj $\left[k[X_0,\ldots,X_n]/(f_1,\ldots,f_m)\right]$, then the k'-rational points of X are given by equivalence classes of non-trivial solutions of the f_j, i.e. by tuples $(x_0,\ldots,x_n)$ in $k'^{n+1} \subseteq \{0\}$ with $f_j(x_0,\ldots,x_n) = 0$ for all j. Another such tuple $(y_0,\ldots,y_n)$ defines the same k'-rational point if and only if there is a $\lambda \in k'^{\times}$ with $y_i = \lambda x_i$ for all i. As usual, we denote the equivalence class of $(x_0,\ldots,x_n)$ by $[x_0 : \ldots : x_n]$ and call this the *homogeneous coordinates* of the corresponding k'-rational point.

Since X, being a variety, is of finite type over k, there is a finite covering $X = \bigcup_{i=1}^n U_i$ of X by affine k-varieties

$$U_i = \mathrm{Spec}\,\left[k[X_1,\ldots,X_{n_i}]/(f_1^{(i)},\ldots,f_{m_i}^{(i)})\right].$$

Let $P : \mathrm{Spec}\,k' \to X$ be a k'-rational point of X. Since $\mathrm{Spec}\,k'$ consists of only one point, P factors over $U_i \hookrightarrow X$ for a suitable i and hence defines a k'-rational point of U_i with affine coordinates $(x_1,\ldots,x_{n_i}) \in k'^{n_i}$.

But k' is a *finite* field, so that there are only finitely many possibilities for such coordinates for each of the finitely many i.

This proves that there are only finitely many k'-rational points of X and shows that the following definition makes sense:

Definition 2.2   *(Zeta function)*

(1) Let X be a k-variety. For a positive natural number i, let $\nu_X^{(i)} \in \mathbb{N}_0$ denote the number $\#X(\mathbb{F}_{q^i})$ of $\mathbb{F}_{q^i}$-rational points of X. Using the numbers $\nu_X^{(i)}$, the *zeta function of X* is the formal power series

$$\zeta(X,T) := \exp\left(\sum_{i=1}^{\infty} \frac{\nu_X^{(i)}}{i} T^i\right) \in \mathbb{Q}[[T]].$$

(2) Let $n \in \mathbb{N}_+$ be a positive natural number, let $P \in k[X_0,\ldots,X_n]$ be a non-constant homogeneous polynomial, and let $X := X(P) := \mathrm{Proj}\,\left[k[X_0,\ldots,X_n]/P\right]$ be the hypersurface in $\mathbb{P}_k^n$ defined by P. Then we define the *zeta function of P*, $\zeta(P,T)$, as the zeta function of X.

Lemma 2.1 *Let X be a k-variety. Denote the set of closed points of X by X^0, and for $x \in X^0$ put $\deg(x) := [\kappa(x) : k]$. Then*

$$\zeta(X,T) = \prod_{x \in X^0} \frac{1}{1 - T^{\deg(x)}}.$$

In particular we see that $\zeta(X,T) \in \mathbb{Z}[[T]]$.

Proof. For $n \in \mathbb{N}_+$, denote the (finite!) number of closed points of X of degree n by d_n. Then

$$\prod_{x \in X^0} \frac{1}{1 - T^{\deg(x)}} = \prod_{n=1}^{\infty} \prod_{\substack{x \in X^0 \\ \deg(x) = n}} \frac{1}{1 - T^n} = \prod_{n=1}^{\infty} \left(\frac{1}{1 - T^n} \right)^{d_n},$$

which implies

$$\log \left(\prod_{x \in X^0} \frac{1}{1 - T^{\deg(x)}} \right) = \sum_{n=1}^{\infty} d_n \cdot \log \left(\frac{1}{1 - T^n} \right)$$

$$= \sum_{n=1}^{\infty} d_n \cdot \sum_{j=1}^{\infty} \frac{T^{jn}}{j} = \sum_{n,j=1}^{\infty} d_n n \frac{T^{jn}}{jn} = \sum_{i=1}^{\infty} \left(\sum_{n|i} d_n n \right) \frac{T^i}{i}.$$

Since every $x \in X^0$ with $\deg(x) = n$ induces n different $\mathbb{F}_{q^n}$-rational points of X, one for each k-automorphism of $\mathbb{F}_{q^n}$, we have $\sum_{n|i} d_n n = \nu_X^{(i)}$, and the proof is complete. $\qquad\qquad\square$

Remark 2.1 *In [Wei49], André Weil formulated the following famous conjectures concerning the zeta function of a smooth, projective, r-dimensional k-variety X (compare [Mil80, p286]):*

(1) (Rationality): $\zeta(X,T)$ is rational, i.e. an element of $\mathbb{Q}(T) \cap \mathbb{Q}[[T]]$.

(2) (Functional equation): The zeta function satisfies the functional equation

$$\zeta \left(X, \frac{1}{q^r T} \right) = \pm q^{\frac{r\chi}{2}} \cdot T^\chi \cdot \zeta(X,T)$$

with χ equal to the Euler-Poincaré characteristic of X (intersection number of the diagonal $\Delta \subseteq X \times X$ with itself).

(3) (Riemann hypothesis): *We have*

$$\zeta(X,T) = \frac{P_1(T) \cdot P_3(T) \cdot \ldots \cdot P_{2r-1}(T)}{P_0(T) \cdot P_2(T) \cdot \ldots \cdot P_{2r}(T)}$$

with $P_0(t) = 1 - T$, $P_{2r}(t) = 1 - q^r T$ *and, for* $1 \le d \le 2r - 1$:

$$P_d(T) = \prod_{i=1}^{\beta_d} (1 - \alpha_{d,i} T)$$

where the $\alpha_{d,i}$ *are algebraic integers of absolute value* $q^{\frac{d}{2}}$ *with respect to every embedding* $\bar{\mathbb{Q}} \hookrightarrow \mathbb{C}$.

(4) (Base change): *If* X *is the reduction of a smooth projective variety* Y *over a number field, then the numbers* β_d *are the Betti numbers of* $Y_{\mathbb{C}}$.

The struggle to prove these conjectures was one of the main motivations for the development of étale and l-adic cohomology by Grothendieck and his coworkers; the most difficult conjecture, namely the Riemann hypothesis, was finally proved by Deligne in his 1974 paper [Del74].

In chapter five, we are going to explain how the Weil conjectures can be proved using l-adic cohomology.

Chapter 3

Galois descent

Throughout this chapter, let K/k be an arbitrary Galois extension of fields with Galois group $G := \mathrm{Gal}(K/k)$.

Galois descent is a common principle in algebra and algebraic geometry that applies in situations where we have the notion of "extending coefficients" from "objects defined over k" to "objects defined over K". If in such a situation it is possible to decide whether an object defined over K is already defined over k, we say that "Galois descent holds".

Of course we will make all this precise later and give an axiomatization of the kind of situation which will be appropriate for all applications in this book, but before we do that, we want to give a few examples:

Example 3.1

(1) The most basic example comes from classical Galois theory itself: The field inclusion $k \hookrightarrow K$ allows us to view elements of k as elements of K. Galois descent clearly holds here, because according to the main theorem of Galois theory, an element $x \in K$ comes from k if and only if $\varphi(x) = x$ for all elements of G.

(2) Let V and W be vector spaces over k, and define the K-vector spaces $V_K := V \otimes_k K$ and $W_K := W \otimes_k K$. Then G acts on V_K and W_K by $^s(v \otimes x) := v \otimes s(x)$ — note that this is only a k-linear action, not a K-linear one. If $f : V_K \to W_K$ is a K-linear map and $s \in G$, we can define another K-linear map $^s f : V_K \to W_K$ by setting $(^s f)(v) := {}^s\big[f({}^{s^{-1}}v)\big]$. Then f comes from k, i.e. is of the form $g \otimes 1_K$ for a k-linear map

15

$g : V \to W$, if and only if ${}^s f = f$ for all $s \in G$. So we have Galois descent in this situation.

(3) Let L be an arbitrary field, let V and W be two finite dimensional vector spaces over L, and let $\varphi : G_k \to \mathrm{Aut}_L(V)$ and $\psi : G_k \to \mathrm{Aut}_L(W)$ be two representations of the absolute Galois group G_k of k. By restricting φ and ψ to the subgroup G_K of G_k, we can also consider those as representations of G_K. Now let $f : V \to W$ be a G_K-equivariant morphism, i.e. an L-linear map satisfying $f[\varphi(t)(x)] = \psi(t)[f(x)]$ for all $t \in G_K$. If s is an element of G, represented by $\tilde{s} \in G_k$, we can define another G_K-equivariant morphism ${}^s f : V \to W$ by setting ${}^s f(x) := \psi(\tilde{s})[f(\varphi(\tilde{s}^{-1})(x))]$ (this definition is independent of the choice of $\tilde{s}$). Then it turns out that f is "defined over k", i.e. f is not only G_K-equivariant but also G_k-equivariant, if and only if ${}^s f = f$ for all $s \in G$. So Galois descent holds in this situation as well.

What do all these examples have in common? — In each case, we have a G-action on the objects in question (elements of K, K-linear maps from V_K to W_K respectively G_K-equivariant maps between two representations), and an object comes from k if and only if it is fix under this action.

We now give a formal definition of this kind of situation, but we restrict our attention to the Galois descent of *isomorphisms*, because that is all we will need in our applications:

Definition 3.1   A *coefficient extension (from k to K)* consists of two categories $\mathcal{C}_k$ and $\mathcal{C}_K$, a covariant functor $F : \mathcal{C}_k \to \mathcal{C}_K$ and for all $Y, Z \in \mathrm{Ob}(\mathcal{C}_k)$ a G-action (from the left) on $\mathrm{Iso}_{\mathcal{C}_K}(FY, FZ)$ (the set of isomorphisms between FY and FZ in $\mathcal{C}_K$), so that the following two conditions are satisfied:

(CE1):   The action is compatible with compositions, i.e. for objects $X, Y, Z \in \mathrm{Ob}(\mathcal{C}_k)$, isomorphisms $X \xrightarrow{g} Y$ and $Y \xrightarrow{f} Z$ and an element $s \in G$, we have:

$$ {}^s(fg) = {}^s f {}^s g. $$

(CE2): Exactly those isomorphisms that come from $\mathcal{C}_k$ are fix under the action of G, i.e. for objects $Y, Z \in \mathrm{Ob}(\mathcal{C}_k)$ we have:

$$ \mathrm{Im}\left(\mathrm{Iso}_{\mathcal{C}_k}(Y, Z) \xrightarrow{F} \mathrm{Iso}_{\mathcal{C}_K}(FY, FZ)\right) = \left[\mathrm{Iso}_{\mathcal{C}_K}(FY, FZ)\right]^G. $$

Often,we do not have to restrict ourselves to isomorphisms but can consider *all* morphisms. The following lemma gives the connection:

Lemma 3.1 *Let $\mathcal{C}_k$, $\mathcal{C}_K$ and $F : \mathcal{C}_k \longrightarrow \mathcal{C}_K$ be as in definition 3.1, and for all objects X and Y from $\mathcal{C}_k$ let us have an action of G on $Mor_K(FX, FY)$, such that the conditions (CE1') and (CE2') that we get by replacing "isomorphism" by "morphism" in (CE1) and (CE2) are satisfied.*
Then we get a coefficient extension after restricting the G-action to isomorphisms.

Proof. This is nearly obvious, we only have to see that the orbit of an isomorphism under the G-action only consists of isomorphisms: If $f : FX \xrightarrow{\sim} FY$ is an isomorphism with inverse $g : FY \xrightarrow{\sim} FX$ and if s is an arbitrary element of G, then

$$^sf \circ {^sg} \overset{\text{CE1'}}{=} {^s(f \circ g)} = {^s1_{FY}} = {^sF1_Y} \overset{\text{CE2'}}{=} 1_{FY}$$

and similarly $^sg \circ {^sf} = 1_{FX}$, so we see that ^sf indeed is an isomorphism with inverse ^sg. $\square$

Now we want to give a series of examples of coefficient extensions, starting with one of the most important ones (for us), the case of varieties over k respectively K. The example is stated for general varieties, even though for our purposes we will only have to consider projective varieties later, and its proof involves more algebraic geometry than most of the rest of the book. The reader not too familiar with these methods will not lose much if he skips the proof of this example.

Example 3.2 Let $F : \mathbf{Var}_k \to \mathbf{Var}_K$ be the functor of base change with K over k

$$X \mapsto X_K := X \times_k K$$
$$(X \xrightarrow{f} Y) \mapsto (X_K \xrightarrow{f \times 1_K} Y_K),$$

and let the action of $s \in G$ on $Mor_{\mathbf{Var}_K}(Y_K, Z_K)$ for varieties Y and Z over

k be defined by the commutativity of the following diagram

$$
\begin{array}{ccc}
Y_K & \xrightarrow{\quad f \quad} & Z_K \\
{\scriptstyle s^*}\big\uparrow{\scriptstyle\wr} & = & {\scriptstyle\wr}\big\uparrow{\scriptstyle s^*} \\
Y_K & \dashrightarrow[{}^{s}\!f] & Z_K
\end{array}
\qquad (3.1)
$$

(note that only the horizontal arrows are morphisms in $\mathbf{Var}_K$ — the vertical arrows are mere morphisms of k-schemes), so that we have

$$
\boxed{{}^{s}\!f = (s^*)^{-1} \circ f \circ (s^*)}. \qquad (3.2)
$$

Because this is not only an important, but also our first example, we want to prove in detail why we really get a coefficient extension in this way.

Proof.   First we have to see that for varieties Y and Z over k, an element s of G and a K-morphism $f : Y_K \to Z_K$, the morphism of schemes ${}^{s}\!f$ defined by (3.2) is really a morphism in $\mathbf{Var}_K$. But this follows immediately from the commutativity of the following diagram of schemes:

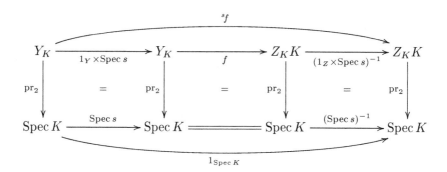

To see that (3.2) defines a G-action on the left, let t be another element of

G; we get:

$$
\begin{aligned}
{}^{1}f &= (1_Z \times 1_{\operatorname{Spec} K})^{-1} \circ f \circ (1_Y \times 1_{\operatorname{Spec} K}) \\
&= 1_{Z_K}^{-1} \circ f \circ 1_{Y_K} \\
&= f,
\end{aligned}
$$

$$
\begin{aligned}
{}^{st}f &= (1_Z \times \operatorname{Spec} [st])^{-1} \circ f \circ (1_Y \times \operatorname{Spec} [st]) \\
&= \Big((1_Z \times \operatorname{Spec} t) \circ (1_Z \times \operatorname{Spec} s)\Big)^{-1} \circ f \circ \Big((1_Y \times \operatorname{Spec} t) \circ (1_Y \times \operatorname{Spec} s)\Big) \\
&= (1_Z \times \operatorname{Spec} s)^{-1} \circ \Big((1_Z \times \operatorname{Spec} t)^{-1} \circ f \circ (1_Y \times t)\Big) \circ (1_Y \times \operatorname{Spec} s) \\
&= (1_Z \times \operatorname{Spec} s)^{-1} \circ {}^{t}f \circ (1_Y \times \operatorname{Spec} s) \\
&= {}^{s}({}^{t}f).
\end{aligned}
$$

To check (CE1'), let X, Y and Z be varieties over k, let $g : X_K \to Y_K$ and $f : Y_K \to Z_K$ be morphisms in $\mathbf{Var}_K$, and let s be an element of G. Then

$$
\begin{aligned}
{}^{s}(fg) &= (1_Z \times \operatorname{Spec} s)^{-1} \circ (fg) \circ (1_X \times \operatorname{Spec} s) \\
&= (1_Z \times s)^{-1} \circ f \circ (1_Y \times s) \\
&\qquad \circ (1_Y \times s)^{-1} \circ g \circ (1_X \times s) \\
&= {}^{s}f \circ {}^{s}g.
\end{aligned}
$$

Finally we have to prove that (CE2') holds, and one inclusion of this is very easy: If $f : X \to Y$ is a morphism of varieties over k, and if s is an element of G, then

$$
\begin{aligned}
{}^{s}(Ff) &= {}^{s}(f \times 1_{\operatorname{Spec} K}) \\
&= (1_Z \times \operatorname{Spec} s)^{-1} \circ (f \times 1_{\operatorname{Spec} K}) \circ (1_Y \times \operatorname{Spec} s) \\
&= f \times 1_{\operatorname{Spec} K} = Ff.
\end{aligned}
$$

So far, our arguments have been quite formal and only used the fact that F was defined by base change with $\operatorname{Spec} K \to \operatorname{Spec} k$ in the category of schemes and that G acts on K respectively $\operatorname{Spec} K$. Whenever we are in a situation like that, diagram (3.1) makes sense and defines a G-action which satisfies (CE1') and the one inclusion of (CE2').

The other inclusion of (CE2'), on the other hand, is *not* purely formal, and to prove it in our case, we will need deeper properties of the categories $\mathbf{Var}_k$ and $\mathbf{Var}_K$ respectively the category of schemes.

It is certainly possible to prove (CE2') completely using affine covers of varieties and polynomial equations over k and K, but we want to give another more conceptual proof here, one that uses the abstract language of "descent" developed in [Gro71, VI,VIII]:

Definition 3.2   Let $p : \mathcal{F} \to \mathcal{S}$ be a (covariant) functor — we also call $\mathcal{F}$ a *category over* $\mathcal{S}$ and p the *projection functor*.

(1) Let S be an object of $\mathcal{S}$. Then we define the category $\mathcal{F}_S$ as the subcategory of $\mathcal{F}$ whose objects are objects $\eta \in \mathrm{Ob}(\mathcal{F})$ with $p\eta = S$ and whose morphisms are morphisms $u : \eta' \to \eta$ in $\mathcal{F}$ with $pu = 1_S$. We call $\mathcal{F}_S$ the *fibre (category) (of p) over* S.

(2) Let $f : T \to S$ be a morphism in $\mathcal{S}$. An f-*morphism* is a morphism α in $\mathcal{F}$ with $p\alpha = f$. If $f = 1_T$, we also speak of a T-*morphism* instead of an 1_T-morphism.

(3) Let $\alpha : \eta \to \xi$ be a morphism in $\mathcal{F}$, and put $T := p\eta$, $S := p\xi$ and $f := p\alpha$ (so that α is an f-morphism according to definition (2)). Then α is called *cartesian* if for all $\eta' \in \mathrm{Ob}(\mathcal{F}_T)$ and all f-morphisms $u : \eta' \to \xi$, there is an unique T-morphism $\bar{u} : \eta' \to \eta$ with $u = \alpha \circ \bar{u}$:

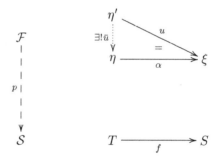

In this case, the object η, together with the f-morphism α, is called the *inverse image of ξ under f* and denoted by $f^*\xi$ — it is unique up to a unique T-morphism.

(4) Let $f : T \to S$ be a morphism in S. If for all objects ξ of $\mathcal{F}_S$, the inverse image of ξ under f exists, we get a covariant functor $f^* : \mathcal{F}_S \to \mathcal{F}_T$ be choosing an inverse image for every object ξ in $\mathcal{F}_S$. We then call that functor the *inverse image functor under f* and say that *the inverse image functor under f in $\mathcal{F}$ exists*.

(5) We say that $\mathcal{F}/\mathcal{S}$ is a *fibred category* (or call p a *fibre functor*) if for all morphisms f in $\mathcal{S}$, the inverse image functor under f in $\mathcal{F}$ exists, and if the composite of two cartesian morphisms in $\mathcal{F}$ is again cartesian.

(6) If p is a fibre functor and if fibre products exist in $\mathcal{S}$, a morphism $f : T \to S$ in $\mathcal{S}$ is called a *morphism of $\mathcal{F}$-descent* if for all objects ξ and ξ' of $\mathcal{F}_S$, the following sequence of sets is exact:

$$\mathrm{Mor}_{\mathcal{F}_S}(\xi, \xi') \xrightarrow{f^*} \mathrm{Mor}_{\mathcal{F}_T}(f^*\xi, f^*\xi') \underset{\mathrm{pr}_2^*}{\overset{\mathrm{pr}_1^*}{\rightrightarrows}} \mathrm{Mor}_{\mathcal{F}_{T \times_S T}}(g^*\xi, g^*\xi'),$$

where g, pr_1 and pr_2 are defined as follows:

$$
\begin{array}{ccc}
T \times_S T & \xrightarrow{\mathrm{pr}_2} & T \\
{\scriptstyle \mathrm{pr}_1}\downarrow & {\scriptstyle g}\searrow & \downarrow{\scriptstyle f} \\
T & \xrightarrow[f]{} & S
\end{array}
$$

For example, let $\mathcal{S}$ be any category, and let $\mathcal{F}$ be the category of morphisms in $\mathcal{S}$. So an object of $\mathcal{F}$ is a morphism in $\mathcal{S}$, and a morphism between objects $f : X \to Y$ and $f' : X' \to Y'$ in $\mathcal{F}$ is a pair of morphisms $\alpha : X \to X'$ and $\beta : Y \to Y'$ in $\mathcal{S}$ such that

$$
\begin{array}{ccc}
X & \xrightarrow{f} & Y \\
{\scriptstyle \alpha}\downarrow & & \downarrow{\scriptstyle \beta} \\
X' & \xrightarrow[f']{} & Y'
\end{array}
\qquad (3.3)
$$

commutes.

$\mathcal{F}$ becomes a category over $\mathcal{S}$ if we define $p : \mathcal{F} \to \mathcal{S}$ by sending an object $f : X \to Y$ of $\mathcal{F}$ to the target Y in $\mathcal{S}$.

It is immediately clear from the definition that a morphism $(\alpha, \beta) : f \to f'$ in $\mathcal{F}$ is cartesian if and only if the diagram (3.3) is cartesian in the usual sense. This means that inverse images exist for all morphisms in $\mathcal{S}$ if and only if all fibred products exist in $\mathcal{S}$, and in this case $\mathcal{F}/\mathcal{S}$ becomes a fibred category.

In particular, for $\mathcal{S}$ we can take the category **Sch** of schemes; as fibred products exist in **Sch**, we thus get a fibred category $\mathcal{F}/\textbf{Sch}$, and we have the following theorem of Grothendieck:

Theorem 3.1   *([Gro71, VIII.5.2])*
Every faithfully flat and quasicompact morphism of schemes is a morphism

of $\mathcal{F}$-descent.

What does that mean in our situation? — First we note that f : $\operatorname{Spec} K \to \operatorname{Spec} k$ is faithfully flat and quasicompact[1], so we can apply theorem 3.1 and see that f is a morphism of $\mathcal{F}$-descent.

Now let Y and Z be varieties over k, and let $\alpha : Y_K \to Z_K$ be a morphism in $\mathbf{Var}_K$.

Let us first assume that K/k is a *finite* Galois extension. We claim that in this case, we have a canonical ring isomorphism $K \otimes_k K \xrightarrow{\gamma} \prod_{s \in G} K$, given by sending $a \otimes b$ to $(a \cdot {}^s b)_{s \in G}$: Indeed, let $x \in K$ be a primitive element, and let $F(X) \in k[X]$ be the minimal polynomial of x (so that we have $k[X]/F(X) \xrightarrow{\sim} K$ be sending X to x). Then we can compute $K \otimes_k K$:

$$K \otimes_k K \cong K \otimes_k [k[X]/F(X)] \cong K[X]/F(X) =$$
$$K[X]/\prod_{s \in G}(X - {}^s x) \cong \prod_{s \in G} K[X]/(X - {}^s x) \cong \prod_{s \in G} K$$

and define γ as the composition. To see that $\gamma(a \otimes b) = (a \cdot {}^s b)_s$ as claimed, it is enough to check the formula for $b = x$ (because x generates K as a k-algebra), and under the chain of isomorphisms given above, $a \otimes x$ is mapped as follows:

$$a \otimes x \mapsto a \otimes [X \pmod{F(X)}] \mapsto aX \pmod{F(X)} \mapsto$$
$$aX \pmod{\prod(X - {}^s x)} \mapsto (aX \pmod{X - {}^s x})_{s \in G} \mapsto (a^s x)_{s \in X}.$$

So we get for each $s \in G$ the following commutative diagram of rings and ring homomorphisms:

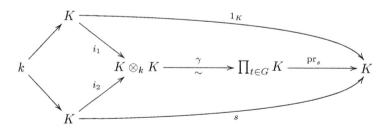

and the following corresponding commutative diagram of (affine) schemes:

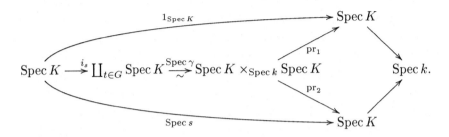

This means that for $s \in G$, the following two diagrams of schemes commute:

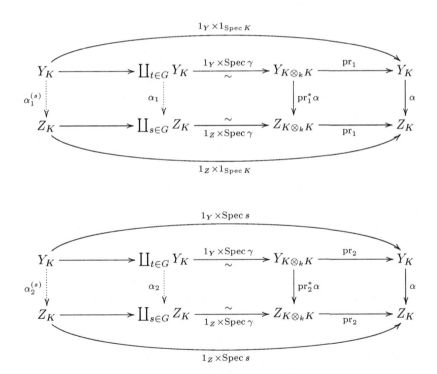

It follows that $\alpha_1 = \coprod_{t \in G} \alpha$ and $\alpha_2 = \coprod_{t \in G} {}^t\alpha$, so that by definition of a morphism of $\mathcal{F}$-descent, we have the following diagram of sets with exact rows (where g denotes the morphism $\operatorname{Spec} k \to K \otimes_k K$):

$$
\begin{array}{ccccc}
\operatorname{Mor}_{\mathcal{F}_k}(Y,Z) & \xrightarrow{\ f^*\ } & \operatorname{Mor}_{\mathcal{F}_K}(f^*Y, f^*Z) & \underset{\operatorname{pr}_2^*}{\overset{\operatorname{pr}_1^*}{\rightrightarrows}} & \operatorname{Mor}_{\mathcal{F}_{K \otimes_k K}}(g^*\xi, g^*\xi') \\
\| & & \| & & \wr \Big\downarrow \gamma_* \\
\operatorname{Mor}_{\mathbf{Var}_k}(Y,Z) & \xrightarrow{\ f^*\ } & \operatorname{Mor}_{\mathbf{Var}_K}(Y_K, Z_K) & \underset{(^s\alpha)_t}{\overset{(\alpha)_t}{\rightrightarrows}} & \prod_{t \in G} \operatorname{Mor}_{\mathbf{Var}_K}(Y_K, Z_K).
\end{array}
$$

So α is mapped to the same element in the rightmost set if and only if it is fix under the action of G, and because of the exactness of the sequence, this is the case if and only if α comes from a morphism $Y \to Z$ in $\mathbf{Var}_k$. — This proves (CE2) in the case that G is finite.

To complete the proof, let G now be arbitrary. Because Y and Z are varieties and therefore of finite type over k, the morphism α is already defined over a field K' that lies between k and K and is *finite* over k — This follows easily from general facts about morphisms between schemes of finite type over an affine projective limit (see [Gro66, 8.8.2(i)]), but can also be seen directly as follows:

The morphism α is given by finitely many morphisms between finitely many affine open subschemes of Y_K and Z_K. Those affine subschemes are defined by finitely many polynomial equations with coefficients in K, and all of the above morphisms are also given by finitely many polynomials with coefficients in K. So as K' we can take the field we get by adjoining all these finitely many coefficients to k, and by taking the normal hull of K', we can further assume that K' is finite and Galois over k with Galois group G'.

If we consider α as a K'-morphism from $Y_{K'}$ to $Z_{K'}$, then by hypothesis α is fix under the action of G', and seeing as G' is finite, it follows from the already proved special case that α is already defined over k. This completes the proof of (CE2') in the general case. $\square$

Now that we have seen a first important example of what we call a coefficient extension, let us look at some more examples:

Example 3.3

(1) Let $\mathbf{Var}'_k$ and $\mathbf{Var}'_K$ be the categories of *geometrically irreducible* varieties over k resp. K with *dominant, rational maps* as morphisms.

Define $\mathbf{Var}'_k \xrightarrow{F} \mathbf{Var}'_K$ as in example 3.2 by base change, and again define the G-action by the commutativity of (3.1). Then with arguments similar to those in example 3.2, one can prove that $\mathbf{Var}'_k \xrightarrow{F} \mathbf{Var}'_K$ is a coefficient extension; but seeing as we will not need this example for our applications, we will not give the details of the proof.

(2) Let $p, q \in \mathbb{N}_0$, and consider the categories $\mathcal{V}^{p,q}_k$ and $\mathcal{V}^{p,q}_K$, whose objects are pairs (V, x), consisting of a finite dimensional k-vector space (respectively K-vector space) V and a tensor x of type (p, q) over V, i.e.

$$x \in T^p_q := V^{\otimes p} \otimes (V^*)^{\otimes q}.$$

Morphisms $(V, x) \xrightarrow{f} (W, y)$ are k-linear (respectively K-linear) *bijections* $V \xrightarrow{f} W$ satisfying

$$\left[f^{\otimes p} \otimes ([f^{-1}]^*)^{\otimes q} \right] (x) = y.$$

Define the functor $F : \mathcal{V}^{p,q}_k \to \mathcal{V}^{p,q}_K$ by sending a pair (V, x) to the pair (V_K, x_K) with $V_K := V \otimes_k K$ and $x_K := x \otimes 1 \in T^p_q(V_K) = T^p_q(V) \otimes_k K$, and a morphism f to $f \otimes 1$, and define the action of $s \in G$ by the commutativity of the following diagram:

$$
\begin{array}{ccc}
(V_K, x_K) & \xrightarrow[\sim]{f} & (W_K, y_K) \\
{\scriptstyle 1 \otimes s} \downarrow \wr & = & \wr \downarrow {\scriptstyle 1 \otimes s} \\
(V_K, x_K) & \xrightarrow[{}^{s}f]{\sim} & (W_K, y_K)
\end{array}
$$

(Note that the vertical arrows are not morphisms in $\mathcal{V}^{p,q}_K$, but merely k-linear maps.)

We thus have:

$$\boxed{{}^s f = (1 \otimes s) \circ f \circ (1 \otimes s)^{-1}}, \tag{3.4}$$

and we claim that $\mathcal{V}^{p,q}_k \xrightarrow{F} \mathcal{V}^{p,q}_K$ is a coefficient extension.

It is easy to see that F and the G-action are well-defined and that (CE1) is satisfied. To prove (CE2), let $f : (V_K, x_K) \xrightarrow{\sim} (W_k, y_K)$ be fixed by G, and let (v_j) and (w_i) be k-bases of V respectively W. Then $(v_j \otimes 1)$ and $(w_i \otimes 1)$ are K-bases of V_K respectively W_K, and if (a_{ij})

denotes the corresponding matrix of f, for $s \in G$ we have:

$$^s f(v_j \otimes 1) \overset{(3.4)}{=} \left[(1 \otimes s) \circ f \circ (1 \otimes s)^{-1} \right] (v_j \otimes 1) = \left[(1 \otimes s) \circ f \right] (v_j \otimes 1)$$

$$= (1 \otimes s) \left(\sum_i w_i \otimes a_{ij} \right) = \sum_i s(a_{ij}) w_i.$$

We therefore get the following formula for the G-action:

$$\boxed{ ^s(a_{ij}) = (^s a_{ij}) }, \tag{3.5}$$

with $^s a_{ij} := s(a_{ij})$. According to the hypothesis, f is fixed by G, so
(3.5) implies that the a_{ij} are fixed as well, so they must be elements of
k. This shows that f is indeed already defined over k.

(3) For $n, r \in \mathbb{N}_+$ consider the categories $\mathcal{F}_k^{n,r}$ and $\mathcal{F}_K^{n,r}$ whose objects are non-zero homogenous polynomials of degree r in $X_1, \ldots, X_n$
over k respectively K and whose morphisms $P \to Q$ are elements
$A = (a_{ij})$ of $\mathrm{GL}(n, k)$ (respectively $\mathrm{GL}(n, K)$), with $Q(AX) :=$
$Q \left(\sum_{j=1}^n a_{1j} X_j, \ldots, \sum_{j=1}^n a_{nj} X_j \right) = P$. Informally speaking, that
means that morphisms are given by a linear change of variables. If
for $F : \mathcal{F}_k^{n,r} \to \mathcal{F}_K^{n,r}$ we take the obvious functor and for a morphism
$(a_{ij}) \in \mathrm{GL}(n, K)$ in $\mathcal{F}_K^{n,r}$ and $s \in G$ we put $^s(a_{ij}) := (^s a_{ij})$, we obviously get a coefficient extension $\mathcal{F}_k^{n,r} \overset{F}{\to} \mathcal{F}_K^{n,r}$.

For example, if $\mathrm{char}(k) \notin \{2, 3\}$, and if we consider $P :=$
$X_1^3 + X_2^3 + X_3^3$ and $Q := X_1^3 + \frac{1}{12} X_2^2 X_3 + \frac{1}{4} X_3^3$ in $\mathcal{F}_k^{3,3}$, then $A :=$
$\left(\begin{smallmatrix} 0 & 0 & 1 \\ 3 & -3 & 0 \\ 1 & 1 & 0 \end{smallmatrix} \right)$ is invertible over k and defines a morphism $P \to Q$ because
of $Q(X_3, 3X_1 - 3X_2, X_1 + X_2) = P$. — This means that P and Q are
"the same" equation modulo the linear change of variables given by A.

(4) For $n, r \in \mathbb{N}_+$ now consider the categories $\widetilde{\mathcal{F}_k^{n,r}}$ and $\widetilde{\mathcal{F}_K^{n,r}}$ whose objects are those of $\mathcal{F}_k^{n,r}$ respectively $\mathcal{F}_K^{n,r}$ but whose morphisms $P \to Q$
are elements of $\mathrm{PGL}(n, k)$ (respectively $\mathrm{PGL}(n, K)$), represented by
regular $n \times n$-matrices $A = (a_{ij})$ with $Q(AX) = \lambda \cdot P$ for a $\lambda \in k^\times$
(respectively $K^\times$). For $F : \widetilde{\mathcal{F}_k^{n,r}} \to \widetilde{\mathcal{F}_K^{n,r}}$ we again take the obvious
functor, and for a morphism $(\overline{a_{ij}}) \in \mathrm{PGL}(n, K)$ in $\widetilde{\mathcal{F}_K^{n,r}}$ and $s \in G$ we
put $^s \overline{(a_{ij})} := \overline{(^s a_{ij})}$ and in this way get another coefficient extension

$$\tilde{F} : \widetilde{\mathcal{F}_k^{n,r}} \to \widetilde{\mathcal{F}_K^{n,r}} :$$

It is clear that $\tilde{F}$ and the G-action are well-defined and that (CE1) is satisfied. To prove (CE2), let $\tilde{A} \in \mathrm{PGL}\,(n, K)$ be fixed by G. Choose a representant A of $\tilde{A}$ which is normalized such that at least one entry is 1 (this is obviously possible). For $s \in G$, by hypothesis there is $\lambda \in K^{\times}$ with ${}^sA = \lambda \cdot A$, but because of ${}^s1 = 1$ we immediately get $\lambda = 1$. This means that for all $s \in G$ we have ${}^sA = A \in \mathrm{GL}\,(n, K)$, which first implies $A \in \mathrm{GL}\,(n, k)$ and then $\tilde{A} \in \mathrm{PGL}\,(n, k)$.

In the explicit example from example (3), $\bar{A} \in \mathrm{PGL}\,(3, k)$ now defines a morphism $P \to Q' := 12Q = 12X_1^3 + X_2^2X_3 + 3X_3^3$ in $\widetilde{\mathcal{F}_k^{3,3}}$.

(5) For a category $\mathcal{C}$ and a group S let $\mathbf{Rep}_{\mathcal{C}}^S$ be the category whose objects are pairs (X, φ) with $X \in \mathrm{Ob}(\mathcal{C})$ and φ an S-action on X, and whose morphisms are S-equivariant morphisms of $\mathcal{C}$.

Let G_k and G_K be the absolute Galois groups of k and K, so that we have $G = G_k/G_K$. Let $\mathcal{C}$ be an arbitrary category, and consider the categories $\mathbf{Rep}_{\mathcal{C}}^{G_k}$ and $\mathbf{Rep}_{\mathcal{C}}^{G_K}$ and the functor $F : \mathbf{Rep}_{\mathcal{C}}^{G_k} \to \mathbf{Rep}_{\mathcal{C}}^{G_K}, (X, \varphi) \mapsto (X, \varphi|G_K)$. For $s \in G$ choose a representant $\tilde{s} \in G_k$ and define the action of s on an isomorphism $(X, \varphi|G_K) \xrightarrow{f} (Y, \psi|G_K)$ by

$$ {}^sf := \psi(\tilde{s}) \circ f \circ \varphi(\tilde{s})^{-1}. \tag{3.6}$$

Then $\mathbf{Rep}_{\mathcal{C}}^{G_k} \xrightarrow{F} \mathbf{Rep}_{\mathcal{C}}^{G_K}$ is a coefficient extension.

(In the special case that $\mathcal{C}$ is the category of vector spaces over a field L, we write $\mathbf{Rep}_{L}^{G_k}$ and $\mathbf{Rep}_{L}^{G_K}$ instead of $\mathbf{Rep}_{\mathcal{C}}^{G_k}$ and $\mathbf{Rep}_{\mathcal{C}}^{G_K}$ — compare example 3.1(3).)

To see that the G-action is well-defined, consider $s \in G$ and $f : (X, \varphi|G_K) \xrightarrow{\sim} (Y, \psi|G_K)$. First of all, we want to show that ${}^sf$ is independent of the choice of the representant $\tilde{s} \in G_k$ of s. So let $\bar{s} \in G_k$ be another representant. Then there is a $t \in G_K$ with $\bar{s} = \tilde{s}t$. Because f is a G_K-equivariant morphism, we have $f \circ \varphi(t^{-1}) = \psi(t^{-1}) \circ f$, and it follows:

$$\psi(\bar{s}) \circ f \circ \varphi(\bar{s})^{-1} = \psi(\tilde{s}t) \circ f \circ \varphi(\tilde{s}t)^{-1}$$

$$= \psi(\tilde{s}) \circ \psi(t) \circ \left(f \circ \varphi(t)^{-1}\right) \circ \varphi(\tilde{s})^{-1}$$
$$= \psi(\tilde{s}) \circ \psi(t) \circ \left(\psi(t)^{-1} \circ f\right) \circ \varphi(\tilde{s})^{-1}$$
$$= \psi(\tilde{s}) \circ f \circ \varphi(\tilde{s})^{-1}.$$

We still have to show that ${}^s f$ is a G_K-equivariant morphism. For that, let t be an arbitrary element of G_K. Because G_K is a normal subgroup of G_k, there is a $t' \in G_K$ with $t\tilde{s} = \tilde{s}t'$. It follows:

$$\psi(t) \circ {}^s f = \psi(t\tilde{s}) \circ f \circ \varphi(\tilde{s}) = \psi(\tilde{s}t') \circ f \circ \varphi(\tilde{s})$$
$$= \psi(\tilde{s}) \circ f \circ \varphi(t'\tilde{s}) = \psi(\tilde{s}) \circ f \circ \varphi(\tilde{s}t)$$
$$= {}^s f \circ \varphi(t).$$

It is easy to see that this defines a G-action which satisfies (CE1), so we only want to show (CE2): Let $f : (X, \varphi | G_K) \xrightarrow{\sim} (Y, \psi | G_K)$ be a morphism in $\mathbf{Rep}_{\mathcal{C}}^{G_K}$ fixed by G. We have to show that f is G_k-equivariant. For this, let $\tilde{s}$ be an arbitrary element of G_k, and let s be its image in G. By hypothesis we have ${}^s f = f$ and therefore get

$$\psi(\tilde{s}) \circ f \circ \varphi(\tilde{s})^{-1} = {}^s f = f \quad \Longrightarrow \quad \psi(\tilde{s}) \circ f = f \circ \varphi(\tilde{s}),$$

and this exactly what we had to prove.

(6) For a category $\mathcal{C}$ let $\mathcal{C}^{\mathrm{Iso}}$ denote the category whose objects are those of $\mathcal{C}$ and whose morphisms are the *isomorphisms* of $\mathcal{C}$. If $\mathcal{C}_k \xrightarrow{F} \mathcal{C}_K$ is a coefficient extension, then clearly $\mathcal{C}_k^{\mathrm{Iso}} \xrightarrow{F | \mathcal{C}_k^{\mathrm{Iso}}} \mathcal{C}_K^{\mathrm{Iso}}$ is one as well.

Now that we have defined coefficient extensions and seen some examples, we want to make coefficient extensions from k to K into a *category*; for that, we have to define the notion of a *morphism* between two coefficient extensions.

Let $F : \mathcal{C}_k \to \mathcal{C}_K$ and $F' : \mathcal{C}_k' \to \mathcal{C}_K'$ be two coefficient extensions from k to K. An obvious candidate for a morphism between F and F' would be a pair of functors $(H_k : \mathcal{C}_k \to \mathcal{C}_k', H_K : \mathcal{C}_K \to \mathcal{C}_K')$ such that the diagram

$$
\begin{array}{ccc}
\mathcal{C}_K & \xdashrightarrow{\;H_K\;} & \mathcal{C}_K' \\
\Big\uparrow{\scriptstyle F} & & \Big\uparrow{\scriptstyle F'} \\
\mathcal{C}_k & \xdashrightarrow[\;H_k\;]{} & \mathcal{C}_k'
\end{array}
\qquad (3.7)
$$

commutes and such that (H_k, H_K) is compatible with the G-actions given by F and F'. What is the latter condition supposed to mean? — Let Y and Z be objects of $\mathcal{C}_k$. Then the coefficient extension F defines a G-action on the set $\mathrm{Iso}_{\mathcal{C}_K}(FY, FZ)$, the coefficient extension F' defines a G-action on the set $\mathrm{Iso}_{\mathcal{C}_{K'}}(F'[H_kY], F'[H_kZ])$, and the functor H_K defines a map

$$\mathrm{Iso}_{\mathcal{C}_K}(FY, FZ) \xrightarrow{H_K} \mathrm{Iso}_{\mathcal{C}_{K'}}(F'[H_kY], F'[H_kZ])$$

between these two G-sets. Now "compatible" means that this map should be G-equivariant for all Y and Z.

Unfortunately, this intuitive definition is a little too restrictive for the applications we have in mind, because for the pairs (H_k, H_K) we want to use, the diagram (3.7) will only commute *up to canonical isomorphisms*. If we take this into account, we get the following definition:

Definition 3.3   Let $F : \mathcal{C}_k \to \mathcal{C}_K$ and $F' : \mathcal{C}_k{}' \to \mathcal{C}_K{}'$ be two coefficient extensions from k to K. A *morphism from F to F'* is a triple (H_k, H_K, h), where $H_k : \mathcal{C}_k \to \mathcal{C}_k{}'$ and $H_K : \mathcal{C}_K \to \mathcal{C}_K{}'$ are covariant functors and $h : H_K F \xrightarrow{\sim} F' H_k$ is an isomorphism of functors,

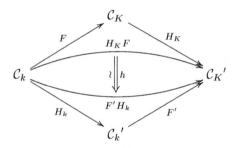

such that for all $Y, Z \in \mathrm{Ob}(\mathcal{C}_k)$ the following composition is G-equivariant:

$$\mathrm{Iso}_{\mathcal{C}_K}(FY, FZ) \xrightarrow{H_K} \mathrm{Iso}_{\mathcal{C}_{K'}}(H_KFY, H_KFZ) \qquad (3.8)$$

$$\wr \downarrow h$$

$$\mathrm{Iso}_{\mathcal{C}_{K'}}(F'[H_kY], F'[H_kZ]).$$

Let $F'' : \mathcal{C}_k{}'' \to \mathcal{C}_K{}''$ be yet another coefficient extension from k to K, and let $\alpha = (H_k, H_K, h)$ and $\beta = (J_k, J_K, j)$ be morphisms from F to F' respectively from F' to F''. Then $\beta\alpha := \beta \circ \alpha$, the *composition of*

β *and* α, is the morphism from F to F'' which is given by the triple $(G_kH_k, G_KH_K, jH_k \circ G_Kh)$:

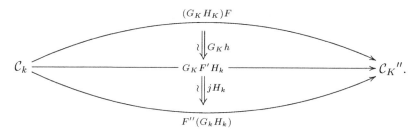

It is not difficult to see that this composition is associative and that $(1_{\mathcal{C}_k}, 1_{\mathcal{C}_K}, 1_F)$ is the identity of $F : \mathcal{C}_k \to \mathcal{C}_K$, so that we indeed get a well defined category with these morphisms and compositions, the *category of coefficient extensions from k to K*.

Example 3.4   Let $n \geq 2$ and $r \geq 1$ be natural numbers.

(1) First we consider the coefficient extensions $\tilde{F} : \widetilde{\mathcal{F}_k^{n,r}} \to \widetilde{\mathcal{F}_K^{n,r}}$ from example 3.3(4) and $F' : \mathbf{Var}_k^{\mathrm{Iso}} \to \mathbf{Var}_K^{\mathrm{Iso}}$ from example 3.2 respectively example 3.3(6).

In order to define a morphism from $\tilde{F}$ to F', we first have to define functors $\tilde{F}_k : \widetilde{\mathcal{F}_k^{n,r}} \to \mathbf{Var}_k^{\mathrm{Iso}}$ and $\tilde{F}_K : \widetilde{\mathcal{F}_K^{n,r}} \to \mathbf{Var}_K^{\mathrm{Iso}}$: For $P \in k[X_1, \ldots, X_n]$ put $\tilde{F}_k P := \mathrm{Proj}\left(k[X_1, \ldots, X_n]/P\right)$ (this is the hypersurface defined by P in the projective space $\mathbb{P}_k^{n-1}$). If $\bar{A} \in \mathrm{PGL}(n,k) : P \to Q$ is a morphism in $\widetilde{\mathcal{F}_k^{n,r}}$, then let $\tilde{F}_k\bar{A}$ be the automorphism of $\mathbb{P}_k^{n-1}$ given by A, restricted to the hypersurface $\tilde{F}_k P$. Obviously $\tilde{F}_k A : \tilde{F}_k P \to \tilde{F}_k Q$ is then a morphism of projective varieties over k, and $\tilde{F}_k$ is a well defined functor from $\widetilde{\mathcal{F}_k^{n,r}}$ to $\mathbf{Var}_k^{\mathrm{Iso}}$. In exactly the same way, replacing k by K, we get a functor $\tilde{F}_K : \widetilde{\mathcal{F}_K^{n,r}} \to \mathbf{Var}_K^{\mathrm{Iso}}$. For each $P \in k[X_i]$ we have a canonical isomorphism

$$\tilde{f}_P : \tilde{F}_K\tilde{F}[P] = \mathrm{Proj}\left(K[X_1, \ldots, X_n]/P\right)$$
$$\xrightarrow{\sim} \mathrm{Proj}\left(k[X_1, \ldots, X_n]/P\right) \times_k K = F'\tilde{F}_k[P],$$

and the $\tilde{f}_P$ induce an isomorphism $\tilde{f} : \tilde{F}_K\tilde{F} \to F'\tilde{F}_k$ of functors. The triple $(\tilde{F}_k, \tilde{F}_K, \tilde{f})$ then defines a morphism from $\tilde{F}$ to F' (it is easy to see that $\tilde{F}_k$, $\tilde{F}_K$ and $\tilde{f}$ fulfill the properties of definition 3.3).

(2) Let $\tilde{F} : \widetilde{\mathcal{F}_k^{n,r}} \to \widetilde{\mathcal{F}_K^{n,r}}$ be as above, and consider in addition to that the coefficient extension $F : \mathcal{F}_k^{n,r} \to \mathcal{F}_K^{n,r}$ from example 3.3(3). We define functors $F_k : \mathcal{F}_k^{n,r} \to \widetilde{\mathcal{F}_k^{n,r}}$ and $F_K : \mathcal{F}_K^{n,r} \to \widetilde{\mathcal{F}_K^{n,r}}$ as follows: We take the identity on objects, and on morphisms we define them by the canonical surjections $\mathrm{GL}(n,k) \twoheadrightarrow \mathrm{PGL}(n,k)$ respectively $\mathrm{GL}(n,K) \twoheadrightarrow \mathrm{PGL}(n,K)$. We have $F_K F = \tilde{F} F_k$, so that the identity is an isomorphism from $F_K F$ to $\tilde{F} F_k$. Then it is easy to see that the triple $(F_k, F_K, 1)$ is a morphism from F to $\tilde{F}$.

So far, we have treated the Galois group G as an abstract group (without topology), but in fact G is a *topological* group, endowed with the Krull-topology.

The following definition takes this topology into account:

Definition 3.4 Let $F : \mathcal{C}_k \to \mathcal{C}_K$ be a coefficient extension. If for all $X, Y \in \mathrm{Ob}(\mathcal{C}_k)$ the G-action on $\mathrm{Iso}_K(FX, FY)$ is *continuous* (where we endow G and $\mathrm{Iso}_K(FX, FY)$ with the Krull-topology respectively the discrete topology), we call that coefficient extension *continuous*.

Example 3.5

(1) We claim that the coefficient extension $F : \mathbf{Var}_k \to \mathbf{Var}_K$ from example 3.2 is continuous. Let Y and Z be varieties over k, and let $f : FY \xrightarrow{\sim} FZ$ be an isomorphism over K. This isomorphism f is given by finitely many polynomials, each of which is given by finitely many coefficients in K. Let $k' \subseteq K$ be the finite field extension of k that we get by adjoining all these finitely many coefficients, and let $G' := \mathrm{Gal}(K/k')$ be the corresponding subgroup of G, which is *open* because k'/k is finite. Then f is already defined over k', which means that all elements of G' leave f fixed. This proves that the G-action on $\mathrm{Iso}_{\mathbf{Var}_K}(FY, FZ)$ is indeed continuous (for this compare lemma 4.1 of the next chapter).

(2) The coefficient extension $\tilde{F} : \widetilde{\mathcal{F}_k^{n,r}} \to \widetilde{\mathcal{F}_K^{n,r}}$ from example 3.3(4) is also continuous: Let P and Q be objects of $\widetilde{\mathcal{F}_k^{n,r}}$, let $\bar{B} \in \mathrm{Iso}_K(\tilde{F}P, \tilde{F}Q) \subseteq \mathrm{PGL}(n,K)$ be an isomorphism in $\widetilde{\mathcal{F}_K^{n,r}}$, and let $B = (b_{ij}) \in \mathrm{GL}(n,K)$

be a lift of $\bar{B}$. Let k' be the finite field extension of $k(b_{ij})$ of k, and let $G' := \mathrm{Gal}(K/k')$ be the corresponding open subgroup of G. Then $\bar{B}$ is fix under the G-action restricted to G', and it follows as before that $\tilde{F}$ is continuous.

(3) The coefficient extension $\mathcal{F}_k^{n,r} \to \mathcal{F}_K^{n,r}$ from example 3.3(3) is continuous; this fact can be proven in exactly the same way as the continuity of $\tilde{F}$.

Example 3.6 The coefficient extension $F : \mathbf{Rep}_L^{G_k} \to \mathbf{Rep}_L^{G_K}$ from example 3.3(5) is in general *not* continuous.

For example, let k be a field, let K be a separable closure of k, let $Z = (V, \varphi)$ be an object of $\mathbf{Rep}_L^{G_k}$, and let us assume that the kernel H of $\varphi : G_k \to \mathrm{Aut}_L(V)$ is *not* open in G_k (to see that this situation is possible, we can take $k = \mathbb{Q}$, $L = \mathbb{Q}_l$ and $Z = \mathbb{Q}_l(1) = (\mathbb{Q}_l, \chi_l)$ — compare definition 5.6 and remark 5.2).

Let $Y := (V, 1)$ be the object of $\mathbf{Rep}_L^{G_k}$ where G_k acts trivially on V. We claim that the G_k-action on $I := \mathrm{Iso}_{\mathbf{Rep}_L^{G_K}}(FY, FZ)$ given by F is not continuous in this case.

Note that an isomorphism from FY to FZ is by definition an L-linear G_K-equivariant map $V \to V$ which is just an L-linear map in this case because $G_K = \{1\}$ since K is separably closed.

So the identity 1_V on V is an element of I, and if the G_k-action on I was continuous, the subgroup H' of G_k of elements fixing 1_V would have to be open.

What is ${}^s 1_V$ for $s \in G_k$? — By definition we have for all $v \in V$ and $s \in G_k$:

$$[{}^s 1_V](v) = \varphi(s)\Big[1_V\big(1(s)^{-1}[v]\big)\Big] = [\varphi(s)](v),$$

so

$$H' = \big\{s \in G_k \,|\, {}^s 1_V = 1_V\big\} = \big\{s \in G_k \,|\, \varphi(s) = 1_V\big\} = H,$$

and we have assumed that H is *not* open in G_k. So the coefficient extension F is not continuous in this case.

Chapter 4

Nonabelian cohomology

Throughout this chapter, let S be a topological group (i.e. a group equipped with a topology that turns both multiplication $S \times S \xrightarrow{\mu} S$ and inverse $S \xrightarrow{i} S$ into continuous maps).

It is well known that the category of discrete S-modules (*abelian* groups A equipped with a left-S-action that is continuous with respect to the discrete topology on A) is an abelian category with enough injective objects and that the functor Γ that maps a discrete S-module A to the abelian group A^S of elements fixed by the action of S is left exact. Thus the abstract machinery of homological algebra is applicable and produces cohomology groups of S with coefficients in A as the right derived functors of Γ: $H^i_{\mathrm{cont}}(S, A) := R^i \Gamma A$.

In the next chapters, we will need a cohomology group $H^1_{\mathrm{cont}}(S, A)$ in the situation where S is the Galois group of a Galois extension K/k, $F : \mathcal{C}_k \to \mathcal{C}_K$ is a coefficient extension, X is an object of $\mathcal{C}_k$ and A is the automorphism group $\mathrm{Aut}_{\mathcal{C}_K}(FX)$.

Unfortunately, such groups of automorphisms will not be abelian in general, so that the construction sketched above does not work, but luckily, also for nonabelian groups A it is possible to define a "nonabelian cohomology" which will behave similar to what we are used to in the abelian case. But there will be great differences as well: For example, we will only define the zeroth and first cohomology, and $H^1_{\mathrm{cont}}(S, A)$ will not be an abelian group anymore but merely a *pointed set*.

In this chapter, we give the definition and basic properties of this nonabelian cohomology. We closely follow Serre's introduction into nonabelian cohomology from [Ser97], but we are slightly more general, considering not only profinite groups S but arbitrary topological ones.

Definition 4.1   An S-*set* is a set with a left-S-action. A *discrete S-set* is an S-set A such that the action $S \times A \to A$ is *continuous* when A is equipped with the discrete topology. (In particular, any S-set is discrete if S carries the discrete topology)

A *morphism of (discrete) S-sets* A and B is a map $A \overset{f}{\to} B$ which is compatible with the S-action, i.e.

$$\forall s \in S \ \forall a \in A : f({}^{s}a) = {}^{s}[f(a)].$$

In this way, we get the *category of (discrete) S-sets* with composition of morphisms given by composition of the underlying maps.

Example 4.1   Let K/k be a Galois extension of fields with Galois group G, and let $F : \mathcal{C}_k \to \mathcal{C}_K$ be a *continuous* coefficient extension from k to K. Then by definition 3.4, for all objects Y and Z of $\mathcal{C}_k$ the set $\mathrm{Iso}_{\mathcal{C}_K}(FY, FZ)$ is a discrete G-set.

We start by proving two additional characterizations for discrete S-sets:

Lemma 4.1   *For an S-set A, the following three statements are equivalent:*

(1) A is a discrete S-set.

(2) For all $a \in A$, the stabilizer $\mathrm{Stab}_S(a) := \{s \in S | {}^{s}a = a\}$ is an open subgroup of S.

(3) It is $A = \bigcup_U A^U$, where the union is taken over all open subgroups U of S and where we set $A^U := \{a \in A | \forall s \in U : {}^{s}a = a\}$.

Proof.

- *(1)$\Rightarrow$(2):* So let $S \times A \overset{\varphi}{\to} A$ be continuous, let $a \in A$, and let $s \in \mathrm{Stab}_S(a)$; we have to show that s is an inner point of $\mathrm{Stab}_S(a)$. Because A discrete, the subset $\{a\} \subseteq A$ is open, so by hypothesis $T := \varphi^{-1}(\{a\})$ is open. Since $u \in \mathrm{Stab}_S(a)$, we have $\varphi(u, a) = {}^{u}a = a$, i.e. $(u, a) \in T$. By definition of the product topology, there must be an open neighbourhood U of s in S with $U \times \{a\} \subseteq T$, i.e.

$$\forall u \in U : {}^u a = \varphi(u, a) = a \qquad \Longrightarrow \qquad s \in U \subseteq \mathrm{Stab}_S(a).$$

It follows that s is indeed an inner point of $\mathrm{Stab}_S(a)$.

- *(2)$\Rightarrow$(3):* The inclusion "$\supseteq$" is trivial, so we only have to prove "$\subseteq$". To this end, let $a \in A$ be an arbitrary element. Then by hypothesis $U := \mathrm{Stab}_S(a)$ is an open subgroup of S, and obviously we have $a \in A^U$.
- *(3)$\Rightarrow$(1):* We have to show that $S \times A \xrightarrow{\varphi} A$ is continuous. A map to a discrete topological space obviously is continuous if and only if it is locally constant, so we have to show that φ is locally constant. For that, let (s, a) be in $S \times A$. By hypothesis, there is an open subgroup U of S with $a \in A^U$. Then $M := sU \times \{a\}$ is an open neighbourhood of (s, a) in $S \times A$, and

$$\forall u \in U : \varphi(su, a) = {}^s({}^u a) = {}^s a,$$

i.e. $\varphi|M$ is constant.

$\square$

Definition 4.2 An *S-group* is a (not necessarily abelian) group A which also is an S-set where S acts via group homomorphisms, i.e.

$$\forall s \in S \; \forall a, b \in A : {}^s(ab) = {}^s a \cdot {}^s b.$$

A *discrete S-group* is an whose underlying S-set is discrete. (In particular, every S-group is discrete if S carries the discrete topology.)

A *morphism of (discrete) S-groups* is a groups homomorphism which is a morphism of the underlying (discrete) S-sets, i.e. an S-equivariant group homomorphism.

In this way, we get the *category of (discrete) S-groups*, where again composition is given by composing the underlying maps.

Example 4.2

(1) An *abelian* group A is a (discrete) S-group if and only if it is a (discrete) S-module in the usual sense.

(2) Let K/k be a Galois extension with Galois group G, let $F : \mathcal{C}_k \to \mathcal{C}_K$ be a coefficient extension from k to K, and let X be an object of $\mathcal{C}_k$. Then by definition of a coefficient extension, the set $A(X) := \mathrm{Aut}_{\mathcal{C}_K}(FX)$ is a G-set, and by (CE1) it is a G-group. If F is *continuous*, then $A(X)$ is a *discrete* G-group (compare example 4.1).

As pointed out above, the nonabelian cohomology $H^1_{\mathrm{cont}}(S, A)$ we are going to define in definition 4.5 will in general not be an abelian group anymore, but merely a pointed set — because of that, we quickly want to remind ourselves of the category of pointed sets and of the notion of exact sequences in that category.

Definition 4.3   The *category of pointed sets* is defined as follows: Objects are pairs (L, x), consisting of a set L and an element $x \in L$, the so called *special point*, and a *morphism of pointed sets* $(L, x) \to (M, y)$ is a map $L \xrightarrow{f} M$ with $f(x) = y$. (Often we will denote a pointed set (L, x) simply by L when the special element x is understood.)

A morphism f of pointed sets is called *injective, surjective* respectively *bijective* if f, considered as a mere map between sets, is injective, surjective respectively bijective.

If (L, x) is a pointed set, then a *pointed subset* of (L, x) is a pointed set (M, y) with $M \subseteq L$ and $x = y$.

If $(L, x) \xrightarrow{f} (M, y)$ is a morphism of pointed sets, then the *kernel of f* is the pointed set

$$\mathrm{Ker}\,(f) := \left(f^{-1}(\{y\}), x \right),$$

and the *image of f* is the pointed set

$$\mathrm{Im}\,(f) := \left(f(L), y \right).$$

A sequence $(L, x) \xrightarrow{f} (M, y) \xrightarrow{g} (N, z)$ of pointed sets is called *exact* if $\mathrm{Ker}\,(g) = \mathrm{Im}\,(f)$. More general, a sequence

$$(L_0, x_0) \xrightarrow{f_0} (L_1, x_1) \xrightarrow{f_1} \ldots \xrightarrow{f_{n-1}} (L_n, x_n) \xrightarrow{f_n} (L_{n+1}, x_{n+1})$$

of pointed sets is called *exact* if for all $1 \leq i \leq n$, the sequence $(L_{i-1}, x_{i-1}) \xrightarrow{f_{i-1}} (L_i, x_i) \xrightarrow{f_i} (L_{i+1}, x_{i+1})$ is exact.

Arbitrary products exist in the category of pointed sets: The product of a family $\{(L_i, x_i)\}_{i \in I}$ of pointed sets is obviously given by $\left(\prod_{i \in I} L_i, (x_i)_{i \in I} \right)$. By $*$ we denote the pointed set $(\{*\}, *)$. Obviously $*$ is both initial and final in the category of pointed sets, i.e. a null object.

We have a natural forgetful functor from the category of groups to the category of pointed sets which sends a group G to the pointed set $(G, 1_G)$. Obviously this functor sends exact sequences to exact sequences.

Remark 4.1 *Let $(L, x) \xrightarrow{f} (M, y)$ be a morphism of pointed sets. Then the sequence $(L, x) \xrightarrow{f} (M, y) \to *$ is exact if and only if f is surjective, and if f is injective, then the sequence $* \to (L, x) \xrightarrow{f} (M, y)$ is exact. — Note that the converse of the latter statement is false in general; put for example $(L, x) := (\{0, 1, 2\}, 0)$, $(M, y) := (\{0, 1\}, 0)$ and $f(0) := 0$, $f(1) := f(2) := 1$. Then $* \to (L, x) \xrightarrow{f} (M, y)$ is exact, but f is obviously not injective.*

Now we can define the nonabelian cohomologies $H^0(S, A)$ and $H^1_{\text{cont}}(S, A)$ which are the subject of this chapter:

Definition 4.4 For an S-group A, the *zeroth cohomology (of S with coefficients in A)* is the group

$$H^0(S, A) := A^S := \{a \in A \,|\, \forall s \in S \,:\, {}^s a = a\}.$$

If $A \xrightarrow{f} B$ is a morphism of S-groups, then obviously $f(A^S) \subseteq B^S$, so that we get an induced group homomorphism

$$H^0(f) := f|A^S : H^0(S, A) \longrightarrow H^0(S, B).$$

In this way, $H^0(S, \_)$ becomes a covariant functor from the category of S-groups to the category of groups.

In what follows, we will often consider the zeroth cohomology as a mere pointed set instead of a group and $H^0(S, \_)$ as a functor to the category of pointed sets.

Definition 4.5   Let A be an S-group. By definition, the set $Z^1(S,A)$ of *1-cocycles of S in A* is the set of maps $s \mapsto a_s$ from S to A, subject to the following *cocycle condition*:

$$\forall s, t \in S: \ a_{st} = a_s{}^s a_t.$$

$Z^1(S,A)$ is a pointed set with special element the *trivial cocycle $s \mapsto 1$*. We define the following equivalence relation on the set of 1-cocycles:

$$(a_s) \sim (a'_s) \ :\Leftrightarrow \ \exists b \in A: \forall s \in S: a'_s = b^{-1} a_s{}^s b.$$

Two equivalent cocycles are called *cohomologous*, and we define the *first cohomology (of S with coefficients in A)* to be the set of equivalence classes $H^1(S,A) := Z^1(S,A)/\sim$ — this is also a pointed set with special point the class of the trivial cocycle.

If $A \xrightarrow{f} B$ is a morphism of S-groups, and if (a_s) and (a'_s) are 1-cocycles of S in A, then one checks easily that (fa_s) and (fa'_s) are 1-cocycles of S in B which are cohomologous if (a_s) and (a'_s) are. Therefore we get an induced morphism

$$H^1(f) : H^1(S,A) \longrightarrow H^1(S,B), \ \ [(a_s)] \mapsto [(fa_s)]$$

of pointed sets. In this way, $H^1(S, \_)$ becomes a covariant functor from the category of S-groups to the category of pointed sets.

Now let A be a *discrete* S-group, and look at the subset $Z^1_{\text{cont}}(S,A) \subseteq Z^1(S,A)$ of *continuous* 1-cocycles of S in A (where we again equip A with the discrete topology). The trivial cocycle is trivially continuous because it is constant, so we can consider $Z^1_{\text{cont}}(S,A)$ as a pointed subset of all 1-cocycles. We denote the pointed set of cohomology classes in $Z^1_{\text{cont}}(S,A)$ by $H^1_{\text{cont}}(S,A)$; we then have an injection of pointed sets

$$H^1_{\text{cont}}(S,A) \hookrightarrow H^1(S,A).$$

(In particular we have $H^1_{\text{cont}}(S,A) = H^1(S,A)$ if S is discrete.)

If B is another discrete S-group, and if $A \xrightarrow{f} B$ is a morphism of discrete S-groups, then $H^1(f)|H^1_{\text{cont}}(S,A)$ clearly factorizes over $H^1_{\text{cont}}(S,B)$ and therefore defines a morphism

$$H^1_{\text{cont}}(f) : H^1_{\text{cont}}(S,A) \longrightarrow H^1_{\text{cont}}(S,B)$$

of pointed sets. In this way, $H^1_{\text{cont}}(S, \_)$ becomes a covariant functor from the category of discrete S-groups to the category of pointed sets.

Remark 4.2 *If A is an abelian S-group and therefore just an S-module in the usual sense (compare example 4.22), then $H^1_{cont}(S, A)$ is just ordinary continuous group cohomology of S with coefficients in A (and so in particular carrying the structure of an abelian group).*

Lemma 4.2 *Let A be an S-group, and let (a_s) be an 1-cocycle of S in A. Then $a_{1_S} = 1_A$.*

Proof. This follows immediately from the cocycle condition:

$$1_A = a_{1_S} \cdot a_{1_S}^{-1} = a_{1_S \cdot 1_S} \cdot a_{1_S}^{-1} = \left(a_{1_S} \cdot {}^{1_S}a_{1_S}\right) \cdot a_{1_S}^{-1}$$
$$= \left(a_{1_S} \cdot a_{1_S}\right) \cdot a_{1_S}^{-1} = a_{1_S}. \quad \square$$

Lemma 4.3 *If S is topologically generated by a single element $\sigma \in S$ (which means that $\langle \sigma \rangle$, the cyclic subgroup of S generated by σ, is dense in S), and if A is a discrete S-group, then a continuous 1-cocycle (a_s) of S in A is uniquely determined by its value a_σ on σ.*

(Note that this situation for example arises when S is the absolute Galois group of a finite field $\mathbb{F}_q$, because in that case S is topologically generated by the Frobenius automorphism $x \mapsto x^q$.)

Proof. Let (a'_s) be another continuous 1-cocycle of S in A with $a'_\sigma = a_\sigma$. Then the cocycle condition and lemma 4.3 imply that (a_s) and (a'_s) agree on $\langle \sigma \rangle$ which by hypothesis is a dense subgroup of S. But two continuous maps which agree on a dense subset of their domain must be equal. $\square$

Lemma 4.4 *Let $S = \langle \sigma \rangle$ be cyclic of order $m \in \mathbb{N}_+$ and discrete, and let A be an S-group. Then*

$$Z^1(S, A) \qquad \longleftrightarrow \qquad \left\{ a \in A \,\middle|\, \prod_{i=0}^{m-1} {}^{\sigma^i}a = 1_A \right\}$$

$$(a_s)_s \qquad \longmapsto \qquad a_\sigma$$

$$\left(\prod_{i=0}^{r-1} {}^{\sigma^i}a\right)_{\sigma^r} \qquad \longleftarrow\!\!\!\mid \qquad a$$

are mutually inverse isomorphisms of pointed sets (with special point 1_A on the righthand side), and two 1-cocycles $(a_s)_s$ and $(a'_s)_s$ are cohomologous if and only if there is $b \in A$ satisfying $a'_\sigma = b^{-1}a_\sigma {}^\sigma b$.

Proof. Let α denote the map "$\rightarrow$", and let β denote the map "$\leftarrow$".

- α *well defined:* First we show the formula

$$\forall (a_s)_s \in Z^1(S, A)\; \forall r \in \mathbb{N}_0\; :\; a_{\sigma^r} = \prod_{i=0}^{r-1} {}^{\sigma^i}a_\sigma. \tag{4.1}$$

by induction on r:

  - $r = 0$: Lemma 4.2.
  - $r \Rightarrow r + 1$: $a_{\sigma^{r+1}} = a_{\sigma^r \cdot \sigma} = a_{\sigma^r} \cdot {}^{\sigma^r}a_\sigma = \left(\prod_{i=0}^{r-1} {}^{\sigma^i}a_\sigma \right) \cdot {}^{\sigma^r}a_\sigma.$

It follows:

$$1_A = a_{1_S} = a_{\sigma^m} \stackrel{(4.1)}{=} \prod_{i=0}^{m-1} {}^{\sigma^i}a_\sigma.$$

- β *well defined:* Let $a \in A$ with $\prod_{i=0}^{m-1} {}^{\sigma^i}a = 1_A$ and $s \in S$ be arbitrary, and choose an $r \in \mathbb{N}_0$ with $s = \sigma^r$. First of all, we want to see that the definition of $\beta(a)_s$ does not depend on the choice of r. So let $r' \in \mathbb{N}_0$ be another natural number with $s = \sigma^{r'}$. Of course we then have $r \equiv r' \bmod m$, and without loss of generality, we can restrict ourselves to the case $r' = r + m$:

$$\prod_{i=0}^{r'-1} {}^{\sigma^i}a = \underbrace{\left(\prod_{i=0}^{m-1} {}^{\sigma^i}a \right)}_{=1} \cdot \left(\prod_{i=m}^{r+m-1} {}^{\sigma^i}a \right) = 1 \cdot \prod_{i=0}^{r-1} {}^{(\sigma^m \cdot \sigma^i)}a = \prod_{i=0}^{r-1} {}^{\sigma^i}a.$$

So $\beta(a)$ is a well defined map from S to A, and we have to check the cocycle condition in order to show that this map is indeed an 1-cocycle of S in A. So consider another element $t \in S$, and choose an $u \in \mathbb{N}_0$

with $t = \sigma^u$. Then we have:

$$\beta(a)_{st} = \beta(a)_{\sigma^{r+u}} = \underbrace{\prod_{i=0}^{r+u-1} {}^{\sigma^i}a = \left(\prod_{i=0}^{r-1} {}^{\sigma^i}a\right)}_{=\beta(a)_s} \cdot \left(\prod_{i=r}^{r+u-1} {}^{\sigma^i}a\right)$$

$$= \beta(a)_s \cdot {}^{\sigma^r}\underbrace{\left(\prod_{i=0}^{u-1} {}^{\sigma^i}a\right)}_{=\beta(a)_t} = \beta(a)_s \cdot {}^s\beta(a)_t.$$

- $\alpha \circ \beta = id$: Obvious.
- $\beta \circ \alpha = id$: Formula (4.1).
- *cohomologous:* The condition is clearly necessary by definition of "co-homologous". To see that it is sufficient, let $(a_s)_s$ and $(a'_s)_s$ be two 1-cocycles with $a'_\sigma = b^{-1}a_\sigma{}^\sigma b$ for a suitable $b \in A$. Then for every $r \in \mathbb{N}_0$ we have:

$$a'_{\sigma^r} \overset{(4.1)}{=} \prod_{i=0}^{r-1} {}^{\sigma^i}a'_\sigma = \prod_{i=0}^{r-1} {}^{\sigma^i}(b^{-1}a_\sigma{}^\sigma b) = \prod_{i=0}^{r-1} {}^{\sigma^i}(b^{-1}){}^{\sigma^i}a_\sigma{}^{\sigma^{i+1}}b$$

$$= b^{-1} \cdot \left(\prod_{i=0}^{r-1} {}^{\sigma^i}a_\sigma\right) \cdot {}^{\sigma^r}b \overset{(4.1)}{=} b^{-1}a_{\sigma^r}{}^{\sigma^r}b,$$

so the two cocycles are indeed cohomologous. $\qquad\square$

In the abelian setting, the cohomology functors are *additive*, i.e. they commute with finite direct sums (which are the same as finite products in abelian categories), and we now want to show that the nonabelian co-homologies $H^0(S, A)$ and $H^1_{\text{cont}}(S, A)$ are compatible with finite products, too.

Lemma 4.5 *Let $r \in \mathbb{N}_+$ be a positive natural number, and let $A_1, \ldots, A_r$ be discrete S-groups.*

(1) The direct product $\prod_{i=1}^{r} A_i$ becomes a discrete S-group via

$$^s(a_1, \ldots, a_r) := ({}^s a_1, \ldots, {}^s a_r) \quad \text{for } s \in S,$$

and together with the obvious morphisms $p_i : \prod_{j=1}^{r} A_j \to A_i$, it becomes a product of $A_1, \ldots, A_r$ in the category of discrete S-groups.

(2) The induced morphisms

$$\iota_0 := \prod H^0(p_i) \quad : H^0(S, \prod_{i=1}^r A_i) \longrightarrow \prod_{i=1}^r H^0(S, A_i) \quad and$$

$$\iota_1 := \prod H^1_{cont}(p_i) : H^1_{cont}(S, \prod_{i=1}^r A_i) \longrightarrow \prod_{i=1}^r H^1_{cont}(S, A_i)$$

are isomorphisms of pointed sets.

(3) For each $i \in \{1, \ldots, r\}$ let B_i be a discrete S-group and $A_i \xrightarrow{\varphi_i} B_i$ a morphism of discrete S-groups. Then the following diagrams commute in the category of pointed sets:

$$
\begin{array}{ccc}
H^0(S, \prod_{i=1}^r A_i) & \xrightarrow{\quad H^0(\prod \varphi_i) \quad} & H^0(S, \prod_{i=1}^r B_i) \\
\iota_0 \downarrow \wr & = & \wr \downarrow \iota_0 \\
\prod_{i=1}^r H^0(S, A_i) & \xrightarrow[\quad \prod H^0(\varphi_i) \quad]{} & \prod_{i=1}^r H^0(S, B_i)
\end{array}
$$

and

$$
\begin{array}{ccc}
H^1_{cont}(S, \prod_{i=1}^r A_i) & \xrightarrow{\quad H^1_{cont}(\prod \varphi_i) \quad} & H^1_{cont}(S, \prod_{i=1}^r B_i) \\
\iota_1 \downarrow \wr & = & \wr \downarrow \iota_1 \\
\prod_{i=1}^r H^1_{cont}(S, A_i) & \xrightarrow[\quad \prod H^1_{cont}(\varphi_i) \quad]{} & \prod_{i=1}^r H^1_{cont}(S, B_i).
\end{array}
$$

Proof.

(1) • *S-group:* Trivial.
  • *discrete:* For an arbitrary $a := (a_i) \in \prod A_i$, there are open subgroups $U_1, \ldots, U_r$ of S with $a_i \in A_i^{U_i}$ for $i \in \{1, \ldots, r\}$ according to lemma 4.1(1)$\Rightarrow$(3). But then $U := \bigcap_{i=1}^r U_i$ is an open subgroup

of S, and we clearly have $a \in (\prod A_i)^U$, so the claim follows from lemma 4.1(3)$\Rightarrow$(1).

- *product:* Obvious.

(2)
- ι_0 *bijective:* This is easy, because

$$(a_i)_i \in \left(\prod_{i=1}^r A_i\right)^S \iff \forall s \in S : \underbrace{{}^s(a_1,\ldots,a_r)}_{=({}^sa_1,\ldots,{}^sa_r)} = (a_1,\ldots,a_r)$$

$$\iff \forall s \in S : {}^sa_1 = a_1, \ldots, {}^sa_r = a_r$$

$$\iff a_1 \in A_1^S, \ldots, a_r \in A_r^S.$$

- ι_1 *bijective:* We claim that the inverse to ι_1 is given as follows:

$$\prod_{i=1}^r H^1_{\text{cont}}(S, A_i) \xrightarrow{j} H^1_{\text{cont}}(S, \prod_{i=1}^r A_i)$$

$$\left(({a_s^{(1)}})_s, \ldots, ({a_s^{(r)}})_s\right) \mapsto \left(a_s^{(1)}, \ldots, a_s^{(r)}\right)_s$$

For this, we first check that the images under this map are indeed cocycles. Let $\left(({a_s^{(1)}})_s, \ldots, ({a_s^{(r)}})_s\right) \in \prod_{i=1}^r H^1_{\text{cont}}(S, A_i)$ and $s,t \in S$ be arbitrary. Then we have:

$$\left(a_{st}^{(1)}, \ldots, a_{st}^{(r)}\right) = \left(a_s^{(1)}\,{}^sa_t^{(1)}, \ldots, a_s^{(r)}\,{}^sa_t^{(r)}\right)$$

$$= \left(a_s^{(1)}, \ldots, a_s^{(r)}\right){}^s\!\left(a_t^{(1)}, \ldots, a_t^{(r)}\right),$$

so the cocycle condition holds.

Next we show that j is well defined. Let $\left(({a_s^{(1)}})_s, \ldots, ({a_s^{(r)}})_s\right) \in \prod_{i=1}^r H^1_{\text{cont}}(S, A_i)$ and $b_i \in A_i$ for $i \in \{1,\ldots,r\}$ be arbitrary. We have to show that then the two 1-cocycles

$$\left(a_s^{(1)}, \ldots, a_s^{(r)}\right)_s \quad \text{and} \quad \left(b_1^{-1}a_s^{(1)}\,{}^sb_1, \ldots, b_r^{-1}a_s^{(r)}\,{}^sb_r\right)_s$$

are cohomologous: For an arbitrary $s \in S$, we have:

$$\left(b_1^{-1}a_s^{(1)}\,{}^sb_1, \ldots, b_r^{-1}a_s^{(r)}\,{}^sb_r\right)_s$$

$$= (b_1,\ldots,b_r)^{-1}\left(a_s^{(1)}, \ldots, a_s^{(r)}\right){}^s(b_1,\ldots,b_r),$$

and we see that j indeed is a well defined morphism of pointed sets.
— That ι_1 and j are mutually inverse to each other is obvious.

(3) This follows immediately from (1) and (2) for categorical reasons. □

Definition 4.6   An *exact sequence of discrete S-groups* is an exact sequence of groups, in which all morphisms are also morphisms of discrete S-groups. If

$$1 \to A \xrightarrow{i} B \xrightarrow{p} C \to 1$$

is a short exact sequence of discrete S-groups, then $iA \trianglelefteq B$ is an S-*invariant* normal subgroup, i.e. $\forall s \in S : \forall a \in A : {}^s ia \in iA$.

Lemma 4.6   *If B is a discrete S-group, and if $A \trianglelefteq B$ is an S-invariant normal subgroup, then restriction of the S-action on B to A gives A the structure of a discrete S-group, we get a well defined continuous action of S on B/A by setting ${}^s \bar{b} := \overline{{}^s b}$, and the sequence*

$$1 \longrightarrow A \longrightarrow B \longrightarrow B/A \longrightarrow 1$$

is a short exact sequence of discrete S-groups.

Proof.

- *action on A well defined:* This is obvious, because A is invariant.
- *action on A continuous:* This follows from the continuity of the S-action on B and from lemma 4.1(1)⇔(2).
- *action on B/A well defined:* Let $s \in S$ and $\bar{b} \in B/A$ be arbitrary. If $b' \in B$ is another represent of $\bar{b}$, then we have to show that $\overline{{}^s b} = \overline{{}^s b'}$ holds: Because of $\bar{b} = \overline{b'}$, there is an $a \in A$ with $b = b'a$, so we have

$$\overline{{}^s b} = \overline{{}^s (b'a)} = \overline{{}^s b' \, {}^s a} = \overline{{}^s b'} \cdot \overline{{}^s a} \overset{{}^s a \in A}{=} \overline{{}^s (b')}.$$

- *action on B/A continuous:* Let $\bar{b}$ be an arbitrary element of B/A, and let $U := \mathrm{Stab}_S(b)$. Because B is a discrete S-group, U is an open subgroup of S by lemma 4.1(1)⇒(3). Then we trivially have $\bar{b} \in (B/A)^U$, and the continuity of the action follows from lemma 4.1(3)⇒(1).
- *sequence exact:* By definition of the S-actions on A and B/A, the canonical group homomorphisms $A \to B$ and $B \to B/A$ are obviously morphisms of discrete S-groups. So the sequence — which is trivially exact in the category of groups — is an exact sequence of discrete S-groups. □

In the abelian case, a short exact sequences of S-modules gives rise to a "long exact sequence in cohomology" which is one of the most important cohomological tools. We are now going to present its nonabelian counterpart.

Proposition 4.1 (Long exact cohomology sequence)
Let $1 \to A \overset{i}{\to} B \overset{p}{\to} C \to 1$ *be a short exact sequence of discrete S-groups. Then we have the following functorial "connecting" morphism of pointed sets:*

$$\delta : H^0(S,C) \to H^1_{cont}(S,A), \quad pb \mapsto [i^{-1}(b^{-1} \cdot {}^s\!b)].$$

Here "functorial" means the following: If

is a commuting diagram of discrete S-groups with exact rows, then the following diagram of pointed sets commutes:

Using δ, we get the following "long" exact sequence of pointed sets:

$$H^0(S,A) \overset{H^0(i)}{\hookrightarrow} H^0(S,B) \overset{H^0(p)}{\longrightarrow} H^0(S,C) \overset{\delta}{\to}$$
$$H^1_{cont}(S,A) \overset{H^1_{cont}(i)}{\longrightarrow} H^1_{cont}(S,B) \overset{H^1_{cont}(p)}{\longrightarrow} H^1_{cont}(S,C).$$

Proof.

- δ *well defined:* First of all, we want to show that for arbitrary $b \in B$ with $pb \in C^S$ and for arbitrary $s \in S$, the element $b^{-1} \cdot {}^s\!b$ lies in the image of i:

$$p\left(b^{-1} \cdot {}^s b\right) = (pb)^{-1} \cdot {}^s(pb) \quad \overset{pb \in C^S}{=} \quad (pb)^{-1} \cdot pb = 1_C$$
$$\implies \quad b^{-1} \cdot {}^s b \in iA.$$

Next we check that $(b^{-1} \cdot {}^s b)$ satisfies the cocycle condition; for that let s and t be arbitrary elements of S:

$$b^{-1} \cdot {}^{st} b = b^{-1} \cdot \left({}^s b \cdot {}^s(b^{-1})\right) \cdot {}^s({}^t b) = \left(b^{-1} \cdot {}^s b\right) \cdot {}^s\left(b^{-1} \cdot {}^t b\right).$$

Now we show that $(b^{-1} \cdot {}^s b)$ is a *continuous* cocycle: Let U be the subgroup $\mathrm{Stab}_S(b)$ of S which is open according to lemma $4.1(1) \Rightarrow (2)$. Then the coset sU is also open in S, and obviously $(b^{-1} \cdot {}^s b)$ is constant on this set, so it is a locally constant map and consequently continuous.

Finally we prove that the definition of δ does not depend on the choice of a preimage for pb. So let $b' \in B$ be another preimage of pb under p, i.e. $b = b'(ia)$ for an $a \in A$.

$$\forall s \in S \;:\; b^{-1} \cdot {}^s b = (b'ia)^{-1} \cdot {}^s(b'ia) = (ia)^{-1} \cdot \left(b'^{-1} \cdot {}^s b'\right) \cdot {}^s(ia)$$
$$\implies \quad (b^{-1} \cdot {}^s b) \sim (b'^{-1} \cdot {}^s b').$$

So the 1-cocycles defined by b and b' are cohomologous and therefore define the same well defined cohomology class.

- δ *functorial:* For arbitrary $pb \in C^S$, we have

$$\left(\delta \circ H^0(\gamma)\right)(pb) = \delta(\gamma pb) = \delta(p'\beta b) = \left(i'^{-1}([\beta b]^{-1} \cdot {}^s[\beta b])\right)$$
$$= \left(i'^{-1}(\beta[\underbrace{b^{-1} \cdot {}^s b}_{\in iA}])\right) = \left(\alpha[i^{-1}(b^{-1} \cdot {}^s b)]\right) = \left(H^1_{\mathrm{cont}}(\alpha) \circ \delta\right)(pb).$$

- $H^0(i)$ *injective:* Trivial.
- *exactness at* $H^0(S,B)$: Trivial.
- *exactness at* $H^0(S,C)$: For arbitrary $b \in B^S$, we have

$$\delta(pb) = \left(i^{-1}[b^{-1} \cdot {}^s b]\right) \overset{{}^s b = b}{=} \left(i^{-1}[b^{-1} \cdot b]\right) = \left(i^{-1} 1_B\right) = (1_A),$$

so b is mapped to the trivial class under δ. For the converse, let pb be an arbitrary element of C^S which is mapped to the trivial class in $H^1_{\mathrm{cont}}(S,A)$. By definition, this means that $\delta(pb)$ is cohomologous to the trivial 1-cocycle, so that there is an $a \in A$ satisfying

$$\forall s \in S : 1_A = a^{-1} \cdot \left[i^{-1}\left(b^{-1} \cdot {}^s b\right)\right] \cdot {}^s a = i^{-1}\left[(b \cdot ia)^{-1} \cdot {}^s(b \cdot ia)\right]$$
$$\implies \quad \forall s \in S : 1_B = (b \cdot ia)^{-1} \cdot {}^s(b \cdot ia)$$
$$\implies \quad b \cdot ia \in B^S.$$

Because of $pb = p(b \cdot ia)$, this proves the claim.

- *exactness at* $H^1_{cont}(S, A)$: If pb is an arbitrary element of C^S, then

$$H^1_{\text{cont}}(i)[\delta(pb)] = [b^{-1} \cdot {}^s b] = [b^{-1} \cdot 1_B \cdot {}^s b] = [1_B] \in H^1_{\text{cont}}(S, B).$$

If conversely (a_s) is a continuous 1-cocycle of S in A such that (ia_s) is cohomologous to the trivial 1-cocycle of S in B, then there is a $b \in B$ with

$$\forall s \in S : ia_s = b^{-1} \cdot 1_B \cdot {}^s b = b^{-1} \cdot {}^s b.$$

This first shows

$$\forall s \in S : {}^s b = b \cdot ia_s \in b \cdot iA \quad \implies \quad pb \in C^S$$

and then $[a_s] = \delta(pb)$.

- *exactness at* $H^1_{cont}(S, B)$: If (a_s) is an arbitrary element of $Z^1_{\text{cont}}(S, A)$, then for all $s \in S$ we trivially have $pia_s = 1_C$, i.e. (a_s) is mapped to the trivial class under $H^1_{\text{cont}}(p) \circ H^1_{\text{cont}}(i)$.

Conversely, let (b_s) be a continuous 1-cocycle of S in B such that (pb_s) is cohomologous to the trivial 1-cocycle of S in C. Then there is a $pb \in C$ with

$$\forall s \in S : pb_s = (pb)^{-1} \cdot 1_C \cdot {}^s(pb) = (pb)^{-1} \cdot {}^s(pb) = (pb)^{-1} \cdot p[{}^s b]. \quad (4.2)$$

Now consider the 1-cocycle $(b'_s) := (bb_s {}^s(b^{-1}))$ which is obviously cohomologous to (b_s). Then it follows:

$$\forall s \in S : pb'_s = p[bb_s {}^s(b^{-1})] = pb \cdot pb_s \cdot p[{}^s(b^{-1})]$$
$$\overset{(4.2)}{=} pb \cdot \left((pb)^{-1} \cdot p[{}^s b]\right) \cdot p[{}^s(b^{-1})] = 1_C,$$

which implies that all b'_s lie in iA; this means

$$[b_s] = [b'_s] = H^1_{\text{cont}}(i)[i^{-1}b'_s].$$

□

Corollary 4.1   *Let* $1 \to A \xrightarrow{i} B \xrightarrow{p} C \to 1$ *be a short exact sequence of S-groups. Then we have a well defined functorial morphism of pointed sets*

$$\delta : H^0(S, C) \to H^1(S, A), \;\; pb \mapsto \left[i^{-1}(b^{-1} \cdot {}^s\!b)\right],$$

and the following sequence is exact in the category of pointed sets:

$$H^0(S, A) \xrightarrow{H^0(i)} H^0(S, B) \xrightarrow{H^0(p)} H^0(S, C)$$
$$\xrightarrow{\delta} H^1(S, A) \xrightarrow{H^1(i)} H^1(S, B) \xrightarrow{H^1(p)} H^1(S, C).$$

Proof.   This follows immediately from proposition 4.1 when we equip S with the discrete topology.   □

As mentioned in the introduction, the "wreath product" $\mu_m \wr S_n$ will play a central role in the following chapters – in definition 4.8 we will define wreath products. In particular, every wreath product is a semidirect product, so as a next step we want study the cohomologies $H^0(S, A)$ and $H^1_{\mathrm{cont}}(S, A)$ in the situation where A is a semidirect product. Before we start, let us quickly rehearse the definition of semidirect products:

Definition 4.7   Let T be a group, and let A be a T-group. Then the *semidirect product of A and T* (with respect to the given T-action on A) is the group $A \rtimes T$ whose underlying set is the cartesian product $A \times T$ (whose elements we often write as $a \cdot t$ or even as at instead of (a, t)) and with the group structure defined by

$$[a \cdot t] \cdot [b \cdot u] := (a^t b) \cdot (tu).$$

It is straightforward to check that this indeed defines a group — in particular, the neutral element is $1_A \cdot 1_T$, and the inverse of $a \cdot t$ is ${}^{(t^{-1})}(a^{-1}) \cdot t^{-1}$; in addition to that we have the useful formula

$$\forall t \in T : \; \forall a \in A : \; t \cdot a \cdot t^{-1} = {}^t\!a. \tag{4.3}$$

Note that when T acts *trivially* on A, then the semidirect product $A \rtimes T$ coincides with the ordinary *direct* product $A \times T$.

We have canonical monomorphisms $i_A : A \hookrightarrow A \rtimes T$, $a \mapsto a \cdot 1_T$, and $i_T :$

$T \hookrightarrow A \rtimes T$, $t \mapsto 1_A \cdot t$, and we have a canonical epimorphism $p : A \rtimes T \twoheadrightarrow T$, $a \cdot t \mapsto t$. These induce a short exact sequence of groups

$$1 \longrightarrow A \xrightarrow{i_A} A \rtimes T \xrightarrow{p} T \longrightarrow 1$$

where i_T provides a splitting of p, i.e. $p \circ i_T = \mathrm{id}_T$. So in particular $i_T T$ is a subgroup of $A \rtimes T$ and $i_A A$ is a normal subgroup of $A \rtimes T$ — we often drop i_A and i_T from the notation and consider A and T themselves as (normal) subgroups of their semidirect product.

Conversely, let

$$1 \longrightarrow A \xrightarrow{i_A} B \xrightarrow{p} T \longrightarrow 1$$

be a short exact sequence of groups with a section i_T of p. Then we can define a left-T-action on A by putting

$$\forall t \in T : \ \forall a \in A : \ {}^t a := i_A^{-1} \big[i_T(t) \cdot i_A(a) \cdot i_T(t^{-1}) \big],$$

and we get a group isomorphism

$$A \rtimes T \xrightarrow{\sim} B, \quad a \cdot t \mapsto i_A(a) \cdot i_T(t) \tag{4.4}$$

with inverse

$$B \xrightarrow{\sim} A \rtimes T, \quad b \mapsto \big[i_A^{-1} \big(b \cdot i_T[pb] \big) \big] \cdot [pb].$$

Example 4.3 Let $C_2 = \langle \tau \rangle$ and C_3 be the cyclic groups with two respectively three elements, and define a left C_2-action on C_3 by setting ${}^\tau x := x^{-1}$. Then $C_3 \rtimes C_2$ is actually isomorphic to the symmetric group S_3, and we get an isomorphism of short exact sequence

$$
\begin{array}{ccccccccc}
1 & \longrightarrow & C_3 & \xrightarrow{i_{C_3}} & C_3 \rtimes C_2 & \xrightarrow{p} & C_2 & \longrightarrow & 1 \\
 & & \downarrow \wr & & \downarrow \wr & & \downarrow \wr & & \\
1 & \longrightarrow & A_3 & \xrightarrow[\text{can}]{} & S_3 & \xrightarrow[\text{sgn}]{} & \{\pm 1\} & \longrightarrow & 1.
\end{array}
$$

Corollary 4.2 (to proposition 4.1)
Let T be a group, and let A be a T-group which is also a discrete S-group in

a "compatible" way, by which we mean that the following condition holds:

$$\forall s \in S : \forall t \in T : \forall a \in A : \,^{s}(\,^{t}a) = \,^{t}(\,^{s}a). \tag{4.5}$$

We can turn the semidirect product $A \rtimes T$ into a discrete S-group by setting $\,^{s}(a \cdot t) := \,^{s}a \cdot t$ for all $s \in S$, $a \in A$ and $t \in T$.

If we let S act trivially on T, then $1 \to A \xrightarrow{i_A} A \rtimes T \xrightarrow{p} T \to 1$ is a short exact sequence of discrete S-groups, and the inclusion $i_T : T \hookrightarrow A \rtimes T$ is a morphism of discrete S-groups.

In the long exact cohomology sequence induced by this short exact sequence, the surjections $H^0(p)$ and $H^1_{cont}(p)$ have canonical sections $H^0(i_T)$ respectively $H^1_{cont}(i_T)$, so the morphism $H^0(S, T) \xrightarrow{\delta} H^1_{cont}(S, A)$ is the null morphism (i.e. the map that sends everything to the special element), and the sequence splits into the following two exact sequences of pointed sets:

$$* \longrightarrow H^0(S, A) \xhookrightarrow{\;H^0(i_A)\;} H^0(S, A \rtimes T) \xrightarrow{\;H^0(p)\;} T \longrightarrow *$$
$$H^0(i_T)$$

$$* \longrightarrow H^1_{cont}(S, A) \xrightarrow{\;H^1_{cont}(i_A)\;} H^1_{cont}(S, A \rtimes T) \xrightarrow{\;H^1_{cont}(p)\;} H^1_{cont}(S, T) \longrightarrow *.$$
$$H^1_{cont}(i_T)$$

Proof.

- $A \rtimes T$ discrete S-group: Consider arbitrary elements $s, s' \in S$, $a, a' \in A$ and $t, t' \in T$. Then

$$\,^{s}(at \cdot a't') = \,^{s}(a\,^{t}a' \cdot tt') = \,^{s}a\,^{s}(\,^{t}a')tt' \overset{(4.5)}{=} \,^{s}a\,^{t}(\,^{s}a')tt'$$
$$= \,^{s}a\,^{t}\,^{s}a't' = \,^{s}(at) \cdot \,^{s}(a't'),$$
$$\,^{ss'}(at) = \,^{ss'}a \cdot t = \,^{s}(\,^{s'}a) \cdot t = \,^{s}(\,^{s'}at) = \,^{s}(\,^{s'}(at)) \text{ and}$$
$$\,^{1}(at) = \,^{1}at = at,$$

so $A \rtimes T$ is an S-group. This S-action is *continuous*, because for arbitrary $a \cdot t \in A \rtimes T$ we have

$$s \in \mathrm{Stab}_S(at) \iff \,^{s}at = at \iff \,^{s}a = a \iff s \in \mathrm{Stab}_S(a),$$

i.e. $\mathrm{Stab}_S(at) = \mathrm{Stab}_S(a)$, and according to lemma 4.1(1)$\Rightarrow$(2) the latter is an open subgroup of S, because A is a *discrete S-group*. Then the claim follows from lemma 4.1(2)$\Rightarrow$(1).

- *exact sequence of discrete S-groups:* Obvious.
- $T \hookrightarrow A \rtimes T$ *morphism of discrete S-groups:* Obvious.
- *exact sequences of pointed sets:* Because of proposition 4.1, we get the following exact sequence of pointed sets:

$$H^0(S,A) \xrightarrow{H^0(i_A)} H^0(S, A \rtimes T) \xrightarrow{H^0(p)} H^0(S,T) \xrightarrow{\delta}$$
$$H^1_{\mathrm{cont}}(S,A) \xrightarrow{H^1_{\mathrm{cont}}(i_A)} H^1_{\mathrm{cont}}(S, A \rtimes T) \xrightarrow{H^1_{\mathrm{cont}}(p)} H^1_{\mathrm{cont}}(S,T),$$

and because S acts trivially on T, we have $H^0(S,T) = T$. The monomorphism $i_T : T \hookrightarrow A \rtimes T$ is a section of p in the category of discrete S-groups. But then by functorality, $H^0(i_T)$ and $H^1_{\mathrm{cont}}(i_T)$ are sections of $H^0(p)$ respectively $H^1_{\mathrm{cont}}(p)$ in the category of pointed sets, so δ is the null morphism in the sequence splits as claimed. $\square$

Lemma 4.7 *Let A and T be groups, and let Γ be a T-set, i.e. a set endowed with a left-T-action. Then A^Γ, the group of maps from Γ to A, becomes a T-group by setting*

$$^t\varphi := \left[\gamma \mapsto \varphi(^{(t^{-1}}\gamma))\right]$$

for all $t \in T$ and $\varphi \in A^\Gamma$.

Proof. First we show that A^Γ is a T-set: Let $\varphi \in A^\Gamma$, $\gamma \in \Gamma$ and $t, u \in T$ be arbitrary. Then

- $(^{1_T}\varphi)(\gamma) = \varphi(^{(1_T^{-1}}\gamma)) = \varphi(\gamma)$,
- $(^{tu}\varphi)(\gamma) = \varphi(^{([tu]^{-1}}\gamma)) = \varphi[^{(u^{-1}})(^{(t^{-1}}\gamma))] = [^u\varphi](^{(t^{-1}}\gamma)) = (^t[^u\varphi])(\gamma)$.

Now we show that T acts by group homomorphisms. Let $\varphi, \psi \in A^\Gamma$, $\gamma \in \Gamma$ and $t \in T$ be arbitrary elements, then we have:

$$[^t(\varphi \cdot \psi)](\gamma) = [\varphi \cdot \psi](^{(t^{-1}}\gamma) = [\varphi(^{(t^{-1}}\gamma)] \cdot [\psi(^{(t^{-1}}\gamma)]$$
$$= [(^t\varphi)(\gamma)] \cdot [(^t\psi)(\gamma)] = [^t\varphi \cdot {}^t\psi](\gamma). \quad \square$$

Definition 4.8   As in lemma 4.7, let A and T be groups, and let Γ be a T-set. Then the *wreath product of A and T (with respect to the T-action on Γ)* is the semidirect product $A \wr_\Gamma T := A \wr T := A^\Gamma \rtimes T$ with respect to the T-action on A^Γ defined in lemma 4.7.

In particular, if T is the symmetric group S_n for an $n \in \mathbb{N}_+$, then T acts naturally on $\Gamma := \{1, 2, \ldots, n\}$, and for any group A we get the wreath product $A \wr S_n$ — all wreath products considered in this book will be of this special type.

In this situation, we identify A^Γ with A^n and write elements of $A \wr S_n$ in the form $(a_1, \ldots, a_n) \cdot \sigma$ with $a_1, \ldots, a_n \in A$ and $\sigma \in S_n$. The S_n-action on $A^\Gamma \cong A^n$ then reads as

$$^\sigma(a_1, \ldots, a_r) = (a_{\sigma^{-1}(1)}, \ldots, a_{\sigma^{-1}(r)}),$$

and for $a, b \in A^r$ and $\sigma, \tau \in S_r$ we consequently get

$$
\begin{aligned}
(a \cdot \sigma)(b \cdot \tau) &= (a\,{}^\sigma b) \cdot (\sigma\tau) & = (a_1 b_{\sigma^{-1}(1)}, \ldots, a_r b_{\sigma^{-1}(r)}) \cdot \sigma\tau \text{ and} \\
(a \cdot \sigma)^{-1} &= {}^{\sigma^{-1}}(a^{-1}) \cdot \sigma^{-1} = (a_{\sigma(1)}^{-1}, \ldots, a_{\sigma(r)}^{-1}) \cdot \sigma^{-1}.
\end{aligned}
$$

Lemma 4.8   *Let $r \in \mathbb{N}_+$ be a positive natural number, and let A be a discrete S-group. Then we can define a continuous S-action on the wreath product $A \wr S_r$ as follows:*

$$^s\!\left((a_i)_i \cdot \sigma\right) := ({}^s a_i)_i \cdot \sigma \qquad \text{(for } s \in S, \ a_1, \ldots, a_r \in A \text{ and } \sigma \in S_r\text{).} \quad (4.6)$$

In this way, $A \wr S_r$ becomes a discrete S-group, and $A^r \trianglelefteq A \wr S_r$ is an S-invariant normal subgroup. We thus get a short exact sequence of discrete S-groups

$$1 \to A^r \xrightarrow{i_{A^r}} A \wr S_r \xrightarrow{p} S_r \to 1. \qquad (4.7)$$

Proof.

- *discrete S-group:* Because S acts continuously on A, by lemma 4.5 it also acts continuously on A^r. For $a \in A^r$, $s \in S$ und $\sigma \in S_r$ we get

$$^s(^\sigma a) = {}^s(a_{\sigma^{-1}(i)})_i = ({}^s a_{\sigma^{-1}(i)})_i = {}^\sigma({}^s a),$$

which means that equation (4.5) from corollary 4.2 is satisfied; according to corollary 4.2 we therefore get an induced continuous S-action on

$A \wr S_r = A^r \rtimes S_r$ which turns $A \wr S_r$ into a discrete S-group, and one checks immediately that that action coincides with the action defined in lemma 4.8.

- *S-invariant normal subgroup:* Corollary 4.2.
- *short exact sequence:* Lemma 4.6.

$\square$

Corollary 4.3   *As in lemma 4.8, let $r \in \mathbb{N}_+$ be a positive natural number, and let A be a discrete S-group. Then we get the following short exact sequences in the category of pointed sets:*

$$* \longrightarrow H^0(S, A^r) \overset{H^0(i_{A^r})}{\hookrightarrow} H^0(S, A \wr S_r) \overset{H^0(p)}{\longrightarrow} S_r \longrightarrow *$$
$$H^0(i_{S_r})$$

$$* \longrightarrow H^1_{cont}(S, A^r) \overset{H^1_{cont}(i_{A^r})}{\longrightarrow} H^1_{cont}(S, A \wr S_r) \overset{H^1_{cont}(p)}{\longrightarrow} H^1_{cont}(S, S_r) \longrightarrow *.$$
$$H^1_{cont}(i_{S_r})$$

Proof.   This follows immediately from lemma 4.8 and corollary 4.2. $\square$

Consider an exact sequence $E \overset{\vartheta}{\to} F \overset{\varphi}{\to} G \overset{\psi}{\to} H$ of abelian groups, and let h be an arbitrary element of H. Then the fibre of ψ over h is either empty or equal to

$$g + \operatorname{Ker}(\psi) = g + \operatorname{Im}(\varphi) \cong g + F/\operatorname{Ker}(\varphi) \cong g + F/\operatorname{Im}(\vartheta)$$

for an arbitrary preimage g of h. So if we know the kernel of φ (which is the fibre over zero), we also know how all the other non-empty fibres "look like"; in particular, if the kernel is finite, all non-empty fibres are also finite and have the same number of elements as the kernel.

For example, if $0 \to A \overset{i}{\to} B \overset{p}{\to} C \to 0$ is a short exact sequence of *abelian* S-groups, then the long exact cohomology sequence is not only a sequence of pointed sets, but of abelian groups.

This implies that for a class $c \in H^1_{cont}(S, C)$, the fibre $\left[H^1_{cont}(p)\right]^{-1}(\{c\})$ over c in $H^1_{cont}(S, B)$ is either empty or equal to $b + \operatorname{Ker}(H^1_{cont}(p)) = b + H^1_{cont}(S, A)/H^0(S, C)$ for an arbitrary element b of the preimage.

If, on the other hand, A, B and C are only S-groups, this is no longer true — in general, there will be no connection between the different fibres whatsoever, and even for the kernel itself it is not clear what $H^1_{\text{cont}}(S, A)/H^0(S, C)$ is supposed to mean.

In the following, we want to show how to overcome these difficulties. First we show how to give a description of the kernel similar to that in the abelian case, then we "twist" the S-action on B to get a new exact sequence in which the fibre we are interested in becomes the kernel.

Lemma 4.9   *Let* $1 \to A \xrightarrow{i} B \xrightarrow{p} C \to 1$ *be a short exact sequence of discrete S-groups. Then we can define a* right-action *of* $H^0(S, C)$ *on* $H^1_{cont}(S, A)$ *as follows:*

$$H^1_{cont}(S, A) \times H^0(S, C) \longrightarrow H^1_{cont}(S, A)$$

$$([a_s] \ , \ pb) \qquad \mapsto \ (a_s)^{pb} := \left[i^{-1}(b^{-1}ia_s \mathfrak{b}) \right].$$

Two classes in $H^1_{cont}(S, A)$ *lie in the same* $H^0(S, C)$-*orbit if and only if they are mapped to the same element in* $H^1_{cont}(S, B)$ *under* $H^1_{cont}(i)$:

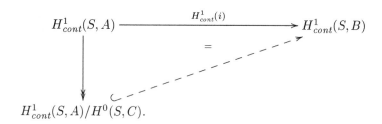

Proof.

- *action well defined:* Let (a_s) be an arbitrary element of $Z^1_{\text{cont}}(S, A)$, and let b be an arbitrary element of B with $pb \in C^S$. First of all, we have to check that for all $s \in S$, the element $b^{-1}ia_s \mathfrak{b}$ lies in the image of i: Because of $pb \in C^S$, there exists an $x_s \in A$ with $\mathfrak{b} = b \cdot ix_s$. So $b^{-1}ia_s \mathfrak{b} = (b^{-1}ia_s b) \cdot x_s \in iA$, because x_s and $b^{-1}ia_s b$ are both elements of iA (the latter holds because iA is a normal subgroup of B). Next we want to check whether $\left(i^{-1}(b^{-1}ia_s b)\right)$ is a 1-cocycle. For that, let s and t be arbitrary elements of S:

$$b^{-1} i a_{st}{}^{st}b = b^{-1} \cdot i[a_s{}^s a_t] \cdot {}^{st}b = b^{-1} i a_s \cdot {}^sb({}^sb)^{-1} \cdot i^s a_t{}^{st}b$$
$$= (b^{-1} i a_s{}^sb) \cdot {}^s(b^{-1} i a_t{}^tb);$$

because i is injective, this proves the cocycle condition.

To prove that the 1-cocycle $\left(i^{-1}(b^{-1}a_s{}^sb)\right)$ is *continuous*, we first choose on open subgroup V of S with $b \in B^V$ according to 4.1(1)$\Rightarrow$(3). If t is an arbitrary element of S, then there is an open neighbourhood V_t of t in S on which the cocycle (a_s) is constant (because (a_s) is a continuous map to a discrete space). Then $U_t := V_t \cap (Vt)$ is also an open neighbourhood of s in S, and $\left(i^{-1}(b^{-1} i a_s{}^sb)\right)$ is obviously constant.

To complete the proof of the action being well defined, we have to show that everything if independent of the choices made. If $b' \in B$ is another element with $pb = pb'$, then there is an $\alpha \in A$ with $b' = b \cdot i\alpha$, and for arbitrary $s \in S$ we have

$$b'^{-1} i a_s{}^sb' = (bi\alpha)^{-1} i a_s{}^sbi\alpha = [i\alpha]^{-1} \cdot (bia_s{}^sb) \cdot {}^s[i\alpha],$$

which shows that the two cocycles $\left(i^{-1}(b'^{-1} i a_s{}^sb')\right)$ and $\left(i^{-1}(b^{-1} i a_s{}^sb)\right)$ are cohomologous.

Finally, let $(a'_s) \in Z^1_{\text{cont}}(S, A)$ be a cocycle which is cohomologous to (a_s). Then there is an $\alpha \in A$ with $a'_s = \alpha^{-1} a_s{}^s\alpha$ for all $s \in S$. So for arbitrary $s \in S$ we get

$$b^{-1} i a'_s{}^sb = b^{-1} \cdot i(\alpha^{-1} a_s{}^s\alpha) \cdot {}^sb = (i\alpha b)^{-1} \cdot i a_s \cdot {}^s(i\alpha b).$$

Because iA is a normal subgroup of B, there is an $\alpha' \in A$ with $i\alpha b = bi\alpha'$, so that we have

$$b^{-1} i a'_s{}^sb = (i\alpha b)^{-1} \cdot i a_s \cdot {}^s(i\alpha b) = (bi\alpha')^{-1} i a_s{}^s(bi\alpha')$$
$$= [i\alpha']^{-1} \cdot \left(b^{-1} i a_s{}^sb\right) \cdot {}^s[i\alpha'].$$

This shows that the cocycles $\left(i^{-1}(b^{-1} i a_s{}^sb)\right)$ and $\left(i^{-1}(b^{-1} i a'_s{}^sb)\right)$ are cohomologous and completes this part of the proof.

• *right-action:* For arbitrary $pb, pc \in C^S$ and $[a_s] \in H^1_{\text{cont}}(S, A)$ we calculate:

$$[a_s]^{1_C} = \left[i^{-1}(1_B^{-1} \cdot ia_s \cdot {}^s 1_B)\right] = [a_s],$$

$$[a_s]^{pb \cdot pc} = [a_s]^{p[bc]} = \left[i^{-1}\left((bc)^{-1} ia_s {}^s(bc)\right)\right]$$
$$= \left[i^{-1}\left(c^{-1} \cdot ii^{-1}(b^{-1} ia_s {}^s b) \cdot {}^s c\right)\right] = \left[[a_s]^{pb}\right]^{pc}.$$

- *orbit:* Look at two classes a and a' in $H^1_{\text{cont}}(S, A)$, represented by continuous 1-cocycles (a_s) respectively (a'_s). Assume first that a and a' lie in the same orbit. There then is a $pb \in C^S$ with $a' = a^{pb}$, so that there is an $\alpha \in A$ with

$$\forall s \in S : a'_s = \alpha^{-1} \cdot i^{-1}(b^{-1} ia_s {}^s b) \cdot {}^s \alpha = (bi\alpha)^{-1} \cdot ia_s \cdot {}^s(bi\alpha).$$

This means that the 1-cocycles (ia_s) and (ia'_s) of S in B are cohomologous.

Conversely suppose $H^1_{\text{cont}}(i)(a) = H^1_{\text{cont}}(i)(a')$ which means that the cocycles (ia_s) and (ia'_s) of S in B are cohomologous. Then there exists a $b \in B$ with

$$\forall s \in S : ia'_s = b^{-1} ia_s {}^s b.$$

This first implies

$$\forall s \in S : {}^s b = [ia_s]^{-1} b[ia'_s] \implies \forall s \in S : p^s b = {}^s[pb] = pb$$
$$\implies pb \in H^0(S, C)$$

and then $a' = a^{pb}$. So a and a' indeed lie in the same orbit.

$\square$

Corollary 4.4 Let $1 \to A \xrightarrow{i} B \xrightarrow{p} C \to 1$ be a short exact sequence of discrete S-groups. Then $H^1_{\text{cont}}(i)$ induces a bijection

$$H^1_{cont}(S, A)/H^0(S, C) \xrightarrow{\sim} Ker\left(H^1_{cont}(S, B) \xrightarrow{H^1_{cont}(p)} H^1_{cont}(S, C)\right).$$

Proof.   This follows immediately from proposition 4.1 and lemma 4.9. $\square$

Lemma 4.10 *Let A be a discrete S-group, and let F be a discrete S-set. We call a left-A-action on F compatible (with the S-action on A and F) if the following condition is satisfied:*

$$\forall s \in S \ \forall a \in A \ \forall f \in F : \ {}^{s}({}^{a}f) = {}^{({}^{s}a)}({}^{s}f).$$

If $a := (a_s) \in Z^1_{cont}(S, A)$ is a continuous 1-cocycle, then

$$ {}^{s\prime}f := a_s({}^{s}f) \tag{4.8}$$

defines a new continuous left-S-action on F, the so-called twisted *action, and we call the resulting discrete S-set the* twist *(of F by a) and denote it by F_a.*

If F is a discrete S-group, and if A acts compatible via group homomorphisms (compare definition 4.2), then F_a is also a discrete S-group.

Proof.

- *left-S-action:* For arbitrary $f \in F$, we have

$$ {}^{(1_S)\prime}f = a_{(1_S)}({}^{(1_S)}f) \overset{\text{lemma } 4.2}{=} {}^{(1_A)}({}^{(1_S)}f) = f,$$

  and for arbitrary $s, t \in S$, we have

$$ {}^{st\prime}f = a_{st}({}^{st}f) = {}^{(a_s \, {}^{s}a_t)}({}^{st}f) = a_s\left({}^{({}^{s}a_t)st}f\right)$$

$$\overset{\text{compatible}}{=} a_s\left({}^{s}\underbrace{\left[a_t({}^{t}f)\right]}_{={}^{t\prime}f}\right) = {}^{s\prime}\left({}^{t\prime}f\right).$$

- *action continuous:* Let f be an arbitrary element of F. Because F is a *discrete* S-set and because of lemma 4.1(1)$\Rightarrow$(3), there is an open subgroup U_1 of S with $f \in F^{U_1}$. On the other hand, a is continuous by hypothesis and accordingly locally constant, so because of lemma 4.2 there is an open subgroup U_2 of S with $a|U_2 = 1_A$. Then $U := U_1 \cap U_2$ is also an open subgroup of S, and obviously ${}^{s\prime}f = f$ for all $s \in U$, so the claim follows from lemma 4.1(3)$\Rightarrow$(1).

- *S-group:* Now let F be a discrete S-group, and assume that A acts on F via group homomorphisms. Let $s \in S$ and $f, g \in F$ be arbitrary. Then we calculate:

$$ {}^{s\prime}(f \cdot g) = a_s({}^{s}(f \cdot g)) = a_s({}^{s}f \cdot {}^{s}g) = a_s({}^{s}f) \cdot a_s({}^{s}g) = {}^{s\prime}f \cdot {}^{s\prime}g.$$

  So in this case F_c really is a discrete S-group. $\square$

Lemma 4.11   *Let B be a discrete S-group, and let $A \trianglelefteq B$ be an S-invariant normal subgroup of B. Then B acts on A via inner automorphisms:*

$$\forall b \in B : \forall a \in A : {}^b a := bab^{-1}.$$

This action is compatible with the S-actions on A and B, and from now on we always want to use this action in this situation.

If $b = (b_s) \in Z_{cont}^1(S, B)$ is a continuous 1-cocycle, then the S-action on the twisted discrete S-group A_b is explicitly given as follows:

$$ {}^{s'}a = b_s \cdot {}^s a \cdot b_s^{-1}. \tag{4.9}$$

(If A lies in the center *of B, then (4.9) immediately shows that the twisted action coincides with the original action, so in this case we have $A = A_b$.)*

In particular, B itself is such an S-invariant normal subgroup of B, so B_b is also defined and equipped with the twisted S-action described by (4.9). It is obvious that the S-action on A_b is simply the restriction of the S-action on B_b; in other words: The group monomorphism $A_b \hookrightarrow B_b$ is again a morphism of discrete S-groups.

Proof.   Let $s \in S$, $a \in A$ and $b \in B$ be arbitrary elements. Then the compatibility is proved by the following calculation:

$$ {}^s({}^b a) = {}^s(bab^{-1}) = {}^sb\, {}^sa \underbrace{{}^s(b^{-1})}_{=({}^sb)^{-1}} = {}^{({}^sb)}({}^sa), $$

and we get equation (4.9) by simply using this compatible action in equation (4.8).   □

What happens if in the situation of lemma 4.11, we have a second continuous 1-cocycle b' which is cohomologous to b? What is the relation between A_b and $A_{b'}$? — In the next lemma, we will show that A_b and $A_{b'}$ are isomorphic (as discrete S-groups, so the isomorphism class of A_b only depends on the cohomology class of b.

Unfortunately, the isomorphism between A_b and $A_{b'}$ is not *canonical*, so that we can not identify the two (compare [Ser97, I.§5.3, p48]).

Lemma 4.12   *Let B be a discrete S-group, let $A \trianglelefteq B$ be an S-invariant normal subgroup of B, and let b and b' be two cohomologous continuous 1-cocycles of S in B. Then the twisted discrete S-groups A_b and A'_b are isomorphic.*

Proof.   Let $b = (b_s)$ and $b' = (b'_s)$. By hypothesis, there is an element $c \in B$ with $b'_s = c^{-1} b_s {}^s c$ for all $s \in S$. Define a map $f : A_b \to A_{b'}$ by $a \mapsto c^{-1} a c$. Then f is obviously an isomorphism of groups — we claim that in addition to that, f is also S-equivariant and thus an isomorphism of discrete S-groups.

Denote the action of S on A_b respectively $A_{b'}$ by $\varphi : S \times A_b \to A_b$ respectively by $\psi : S \times A_{b'} \to A_{b'}$. By lemma 4.11, we have for all $s \in S$ and $a \in A$:

$$\varphi(s, a) = b_s \cdot {}^s a \cdot b_s^{-1},$$
$$\psi(s, a) = b'_s \cdot {}^s a \cdot {b'_s}^{-1} = c^{-1} b_s {}^s c \cdot {}^s a \cdot {}^s(c^{-1}) b_s^{-1} c.$$

So for arbitrary $s \in S$ and $a \in A_b$ we get:

$$\psi(s, f[a]) = \psi(s, c^{-1} a c) = c^{-1} b_s {}^s c \cdot {}^s(c^{-1} a c) \cdot {}^s(c^{-1}) b_s^{-1} c$$
$$= c^{-1} b_s {}^s[cc^{-1}] {}^s a {}^s[cc^{-1}] b_s^{-1} c = c^{-1} \cdot [b_s {}^s b_s^{-1}] \cdot c = f[\varphi(a, s)].$$
$\square$

Because of lemma 4.12, the isomorphism class of a twisted S-group only depends on the cohomology class of the twisting cocycle. Therefore in future, when we are in the situation of lemma 4.11, we will sometimes write $A_{[b]}$ instead of A_b when we are only interested in A_b up to isomorphism.

Lemma 4.13   *Let $B \xrightarrow{\varphi} C$ be a morphism of discrete S-groups, and let $b = (b_s) \in Z^1_{cont}(S, B)$ be a continuous 1-cocycle. Then $\varphi \circ b = (\varphi[b_s])$ is a continuous 1-cocycle of S in C, so the twist $C_{\varphi \circ b}$ is defined by lemma 4.11 — we will simply write C_b instead of $C_{\varphi \circ b}$ when φ is understood.*

In this situation, φ is S-equivariant with respect to the twisted S-actions on B and C, so that we can consider φ as a morphism of discrete S-groups $B_b \xrightarrow{\varphi} C_b$.

Proof.   We only have to show that $B_b \xrightarrow{\varphi} C_b$ is an S-equivariant map. So let $s \in S$ and $b \in B$ be arbitrary elements. We get:

$$\varphi(^{s\prime}b) \overset{(4.9)}{=} \varphi(b_s{}^s b b_s^{-1}) = \varphi(b_s) \cdot \varphi(^s b) \cdot \varphi(b_s^{-1})$$

$$= (\varphi \circ b)_s \cdot {}^s\varphi(b) \cdot (\varphi \circ b)_s^{-1} \overset{(4.9)}{=} {}^{s\prime}\varphi(b). \qquad \Box$$

Corollary 4.5   *Let* $1 \to A \xrightarrow{i} B \xrightarrow{p} C \to 1$ *be a short exact sequence of discrete S-groups, and let* $b \in Z^1_{cont}(S, B)$ *be a continuous 1-cocycle. Then the following sequence is exact in the category of pointed sets:*

$$H^0(S, A_b) \xrightarrow{H^0(i)} H^0(S, B_b) \xrightarrow{H^0(p)} H^0(S, C_b) \xrightarrow{\delta}$$

$$H^1_{cont}(S, A_b) \xrightarrow{H^1_{cont}(i)} H^1_{cont}(S, B_b) \xrightarrow{H^1_{cont}(p)} H^1_{cont}(S, C_b).$$

Proof.   Because of lemma 4.11 and lemma 4.13, the group homomorphisms $A_b \xrightarrow{i} B_b$ respectively $B_b \xrightarrow{p} C_b$ are also morphisms of discrete S-groups, so that we have a short exact sequence of discrete S-groups

$$1 \longrightarrow A_b \xrightarrow{i} B_b \xrightarrow{p} C_b \longrightarrow 1.$$

The claim then follows from proposition 4.1. $\Box$

Lemma 4.14   *Let B be a discrete S-group, and let* $b = (b_s) \in Z^1_{cont}(S, B)$ *be a continuous 1-cocycle. Then the map*

$$t_b : Z^1_{cont}(S, B_b) \longrightarrow Z^1_{cont}(S, B)$$
$$(a_s) \quad \mapsto \quad (a_s \cdot b_s)$$

is a bijection of sets such that two cocycles $a, \tilde{a} \in Z^1_{cont}(S, B)$ *are cohomologous if and only if $t_b(a)$ and $t_b(\tilde{a})$ are cohomologous. Therefore we get an induced bijection*

$$\tau_b : H^1_{cont}(S, B_b) \xrightarrow{\sim} H^1_{cont}(S, B).$$

Under τ_b, the trivial cohomology class is mapped to the class $[b]$.

Proof.

- t_b *well defined:* Let $a = (a_s)$ be an arbitrary continuous 1-cocycle of S in B_b. Then $(a_s \cdot b_s)$ is obviously continuous, because a and b are continuous, so we only have to show that $(a_s \cdot b_s)$ satisfies the cocycle condition. For that, let s and t be arbitrary elements of S. Then

$$[t_b(a)]_{st} = a_{st}b_{st} = a_s{}^{s\prime}a_t b_s{}^s b_t \overset{(4.9)}{=} a_s b_s{}^s a_t b_s^{-1} b_s{}^s b_t$$
$$= a_s b_s{}^s a_t{}^s b_t = a_s b_s{}^s(a_t b_t) = [t_b(a)]_s{}^s[t_b(a)]_t.$$

- t_b *bijective:* We have an obvious candidate for the inverse, namely

$$s_b : Z^1_{\text{cont}}(S, B) \longrightarrow Z^1_{\text{cont}}(S, B_b)$$
$$(a_s) \qquad \mapsto \quad (a_s \cdot b_s^{-1}),$$

and we only have to show that s_b is well defined, which means that we have to check the cocycle condition again (it is obvious that s_b maps continuous cocycles to continuous cocycles). So let $s, t \in S$ and $a = (a_s) \in Z^1_{\text{cont}}(S, B)$ arbitrary. Then we calculate:

$$[s_b(a)]_{st} = a_{st}b_{st}^{-1} = a_s{}^s a_t(b_s{}^s b_t)^{-1} = a_s(b_s^{-1}b_s)^s a_t({}^s b_t)^{-1}b_s^{-1}$$
$$= a_s b_s^{-1}b_s{}^s a_t{}^s(b_t^{-1})b_s^{-1} \overset{(4.9)}{=} a_s b_s^{-1}{}^{s\prime}(a_t b_t^{-1}) = [s_b(a)]_s{}^{s\prime}[s_b(a)]_t.$$

- *respects cohomologous cocycles:* We have to show that t_b and s_b map cohomologous cocycles to cohomologous cocycles. First let $a = (a_s)$ and $\tilde{a} = (\tilde{a}_s)$ be cohomologous continuous 1-cocycles of S in B_b. Then there is a $\beta \in B$ with

$$\forall s \in S : \tilde{a}_s = \beta^{-1}a_s{}^{s\prime}\beta \overset{(4.9)}{=} \beta^{-1}a_s b_s{}^s\beta b_s^{-1}.$$

It follows that

$$[t_b(\tilde{a})]_s = \beta^{-1}a_s b_s{}^s\beta b_s^{-1} \cdot b_s = \beta^{-1} \cdot a_s b_s \cdot {}^s\beta = \beta^{-1} \cdot [t_b(a)]_s \cdot {}^s\beta,$$

which shows that $t_b(a)$ and $t_b(\tilde{a})$ are really cohomologous.

Conversely, let $a = (a_s)$ and $\tilde{a} = (\tilde{a}_s)$ be cohomologous continuous 1-cocycles of S in B. Then there is a $\beta \in B$ with $\tilde{a}_s = \beta^{-1}a_s{}^s\beta$ for all $s \in S$. This implies

$$[s_b(\tilde{a})]_s = \beta^{-1}a_s{}^s\beta \cdot b_s^{-1} = \beta^{-1}a_s \cdot b_s^{-1}b_s \cdot {}^s\beta b_s^{-1}$$
$$= \beta^{-1} \cdot a_s b_s^{-1} \cdot b_s{}^s\beta b_s^{-1} \overset{(4.9)}{=} \beta^{-1} \cdot [s_b(a)]_s \cdot {}^{s\prime}\beta,$$

so $s_b(a)$ and $s_b(\tilde{a})$ are also cohomologous.

• $\tau_b[1_B] = [b]$: This is obvious from the definition of t_b.   □

Lemma 4.15   *Let $B \xrightarrow{\varphi} C$ be a morphism of discrete S-sets, and let $b = (b_s)$ be a continuous 1-cocycle of S in B. Then the following diagram is commutative in the category of sets, but in general* not *in the category of pointed sets:*

$$
\begin{array}{ccc}
H^1_{cont}(S, B) & \xrightarrow{\ H^1_{cont}(\varphi)\ } & H^1_{cont}(S, C) \\
\uparrow \tau_b \wr & = & \wr \uparrow \tau_{\varphi \circ b} \\
H^1_{cont}(S, B_b) & \xrightarrow[\ H^1_{cont}(\varphi)\ ]{} & H^1_{cont}(S, C_b)
\end{array}
$$

Proof.   Let $[\beta] \in H^1_{\text{cont}}(S, B_b)$ be an arbitrary class which is represented by the continuous 1-cocycle $\beta = (\beta_s)$. Then we calculate:

$$
\begin{aligned}
\left(H^1_{\text{cont}}(\varphi) \circ \tau_b\right)[\beta] &= H^1_{\text{cont}}(\varphi)[\beta_s \cdot b_s] = [\varphi(\beta_s b_s)] \\
&= [\varphi(\beta_s) \cdot \varphi(b_s)] = \tau_{\varphi \circ b}[\varphi(\beta_s)] = \left(\tau_{\varphi \circ b} \circ H^1_{\text{cont}}(\varphi)\right)[\beta].
\end{aligned}
$$
          □

Corollary 4.6   *Let $1 \to A \xrightarrow{i} B \xrightarrow{p} C \to 1$ be a short exact sequence of discrete S-groups, and let $b \in Z^1_{cont}(S, B)$ be a continuous 1-cocycle of S in B. Then the following diagram of sets is commutative, and the bottom row is exact in the category of pointed sets:*

$$
\begin{array}{ccccc}
& & H^1_{cont}(S, B) & \xrightarrow{\ H^1_{cont}(p)\ } & H^1_{cont}(S, C) \\
& & \uparrow \tau_b \wr & = & \wr \uparrow \tau_{pb} \\
H^0(S, C_b) \xrightarrow{\ \delta\ } & H^1_{cont}(S, A_b) & \xrightarrow[H^1_{cont}(i)]{} H^1_{cont}(S, B_b) & \xrightarrow[H^1_{cont}(p)]{} & H^1_{cont}(S, C_b).
\end{array}
$$

The map $\tau_b \circ H^1_{cont}(i)$ induces a bijection

$$H^1_{cont}(S, A_b)/H^0(S, C_b) \xrightarrow{\sim} H^1_{cont}(p)^{-1}[pb] \subseteq H^1_{cont}(S, B)$$

$$(a_s) \cdot H^0(S, C_b) \qquad \mapsto \qquad (a_s \cdot b_s).$$

Proof. This follows immediately from corollary 4.4, corollary 4.5 and lemma 4.15. $\qquad\qquad\qquad\qquad\qquad\qquad\qquad\qquad\qquad\qquad\qquad\qquad$ $\square$

Proposition 4.2 *As in corollary 4.2, let T be a group, and let A be a T-group as well as a discrete S-group such that (4.5) holds. So according to corollary 4.2, we have the following short exact sequence of discrete S-groups:*

$$1 \longrightarrow A \xrightarrow{i_A} A \rtimes T \xrightarrow{p} T \longrightarrow 1.$$
$$\overset{\cdots}{\nwarrow}\,{}_{i_T}\,{\cdots}{\nearrow}$$

Let b be a continuous 1-cocycle of S in the semidirect product $A \rtimes T$, and put $t := p \circ b \in Z^1_{cont}(S, T)$ and $\beta := i_T \circ t = 1_T \circ p \circ b \in Z^1_{cont}(S, A \rtimes T)$.

(1) The group homomorphism $i_T : T_\beta \to (A \rtimes T)_\beta$ is a morphism of discrete S-groups, and we have the following short exact sequence of discrete S-groups:

$$1 \longrightarrow A_\beta \xrightarrow{i_A} (A \rtimes T)_\beta \xrightarrow{p} T_\beta \longrightarrow 1. \qquad (4.10)$$
$$\overset{\cdots}{\nwarrow}\,{}_{i_T}\,{\cdots}{\nearrow}$$

(2) The long exact cohomology sequence to the short exact sequence (4.10) splits, because $H^0(p)$ has the section $H^0(i_T)$ because of (1), and we get the following two split short exact sequences, where the first sequence is a sequence in the category of groups, and the second sequence is a sequence in the category of pointed sets.

$$1 \longrightarrow H^0(S, A_\beta) \overset{H^0(i_A)}{\longrightarrow} H^0(S, (A \rtimes T)_\beta) \overset{H^0(p)}{\longrightarrow} H^0(S, T) \longrightarrow 1$$
$$\overset{\cdots}{\nwarrow}\,{}_{H^0(i_T)}\,{\cdots}{\nearrow}$$

$$* \longrightarrow H^1_{cont}(S, A_\beta) \overset{H^1_{cont}(i_A)}{\longrightarrow} H^1_{cont}(S, (A \rtimes T)_\beta) \overset{H^1_{cont}(p)}{\longrightarrow} H^1_{cont}(S, T) \longrightarrow *$$
$$\overset{\cdots}{\nwarrow}\,{}_{H^1_{cont}(i_T)}\,{\cdots}{\nearrow}$$

(3) We have a group isomorphism

$$H^0(S, A_\beta) \rtimes H^0(S, T_\beta) \xrightarrow{\sim} H^0(S, (A \rtimes T)_\beta)$$

$$(a \quad , \quad t) \quad \mapsto \quad i_A(a) \cdot i_T(t).$$

(4) Assume that A is abelian. *Then the twist A_b only depends on the continuous 1-cocycle t of S in T, i.e. $A_\beta = A_b$.*

Proof. Let $b = (a_s \cdot t_s)$ for $a_s \in A$ and $t_s \in T$. So we have $t = (t_s)$ and $\beta = (1 \cdot t_s) =: (\beta_s)$.

(1) We have to check that i_T is S-equivariant with respect to the twisted action. So let $s \in S$ and $\tau \in T$ be arbitrary elements. We get:

$$i_T\big[{}^{s'}\tau\big] \overset{(4.9)}{=} i_T\big[t_s{}^s\tau t_s^{-1}\big] = i_T\big[t_s\tau t_s^{-1}\big] = (1 \cdot t_s)(1 \cdot \tau)(1 \cdot t_s^{-1})$$
$$= \beta_s(1 \cdot \tau)\beta_s^{-1} = \beta_s{}^s(1 \cdot \tau)\beta_s^{-1} = {}^{s'}(i_T[\tau]).$$

(2) This is clear.
(3) This follows immediately from the first short exact sequence of (2) and from (4.4).
(4) Let $s \in S$ and $a \in A$ be arbitrary elements. Then because of (4.9) we have to show $b_s \cdot {}^s a \cdot b_s^{-1} = t_s \cdot {}^s a \cdot t_s^{-1}$:

$$b_s \cdot {}^s a \cdot b_s^{-1} = (a_s \cdot t_s) \cdot {}^s a \cdot (a_s \cdot t_s)^{-1} = (a_s \cdot t_s) \cdot {}^s a \cdot \big({}^{t_s^{-1}}(a_s^{-1}) \cdot t_s^{-1}\big)$$
$$= \Big[a_s \cdot {}^{t_s}({}^s a) \cdot {}^{t_s \cdot t_s^{-1}}(a_s^{-1})\Big] \cdot \big[t_s \cdot t_s^{-1}\big] = \big[a_s \cdot {}^{t_s}({}^s a) \cdot (a_s^{-1})\big]$$
$$\overset{A \text{ abelian}}{=} {}^{t_s}({}^s a) = t_s \cdot {}^s a \cdot t_s^{-1}. \qquad \square$$

Lemma 4.16 *Let A be a group, considered as a discrete S-group with the trivial S-action. Then a continuous 1-cocycle of S in A simply is a continuous group homomorphism from S to A, and two cocycles (a_s) and (a_s') are cohomologous if and only if they are conjugated, i.e. if there is an $b \in A$ such that $a_s' = b^{-1} a_s b$ for all $s \in S$.*

Proof. In this situation, the cocycle condition for a cocycle (a_s) becomes $a_{st} = a_s{}^s a_t = a_s a_t$, but this is precisely the condition for the map $s \mapsto a_s$ to be a group homomorphism.

By definition, two cocycles (a_s) and (a'_s) are cohomologous if and only if there is a $b \in A$ with $a'_s = b^{-1} a_s {}^s b = b^{-1} a_s b$, i.e. if and only if they are conjugated. □

Definition 4.9 Let k be a field. A *finite, commutative, separable k-algebra* (or *fcs-algebra over k* for short) is a commutative k-algebra which is finite and étale over k. Equivalently, an fcs-algebra over k is a k-algebra which is isomorphic (as a k-algebra) to a finite direct product $L_1 \times \ldots \times L_m$ of finite separable field extensions L_i of k. The *degree of an fcs-algebra L over k* is its dimension $\dim_k L$ as a k-vector space. So when $L = L_1 \times \ldots \times L_m$, then the degree of L is $[L_1 : k] + \ldots + [L_m : k]$.

Note that if L is an fcs-algebra of degree n over k and if K is a separable closure of k, then the set $\operatorname{Hom}_k(L, K)$ of k-algebra homomorphisms from L to K has n elements.

Proposition 4.3 *Let k be a field with separable closure K and absolute Galois group $G_k = \operatorname{Gal}(K/k)$, let $n \in \mathbb{N}_+$ be a positive natural number, and let M be a set with n elements. Then there is a canonical isomorphism*

$$\{\text{fcs-algebras of degree } n \text{ over } k\}/k\text{-algebra-isomorphisms}$$
$$\cong \{\text{continuous left-}G_k\text{-actions on } M\}/\text{isomorphisms}.$$

Here G_k and M are endowed with the Krull-topology respectively the discrete topology, and two left-G_k-actions φ, φ' on M are "isomorphic" if there is a bijection $f : M \to M$ such that $\varphi'(s, m) = f^{-1}[\varphi(s, fm)]$ for all $s \in S$ and $m \in M$.

If L is an fcs-algebra of degree n over k, then the corresponding action on the righthand side is given as follows: Put $M' := \operatorname{Hom}_k(L, K)$, choose a bijection $f : M \to M'$ (we have noted in definition 4.9 that M has n elements), and define

$$\begin{aligned} G_k \times M &\longrightarrow M \\ (s \quad , \; m) &\mapsto f^{-1}[s \circ (fm)]. \end{aligned}$$

Proof. [Tam94, II.2.1, p93]. □

Corollary 4.7   *Let $n \in \mathbb{N}^+$ be a positive natural number, consider the symmetric group S_n as a discrete S-group with the trivial action, and let M be a set with n elements. Then we have the following canonical isomorphism of pointed sets:*

$$H^1_{cont}(S, S_n) \cong \{continuous\ left\text{-}S\text{-}actions\ on\ M\}/isomorphisms. \quad (4.11)$$

So if k is a field with separable closure K and if S is the absolute Galois group $G_k = Gal(K/k)$, then proposition 4.3 shows that there is a canonical isomorphism of pointed sets

$$H^1_{cont}(G_k, S_n) \cong \{fcs\text{-}algebras\ of\ degree\ n\ over\ k\}/k\text{-}isomorphisms.$$
$$(4.12)$$

If L is an fcs-algebra of degree n over k, and if we take $M := Hom_{k\text{-}alg}(L, K)$, then the cohomology class corresponding to L via (4.12) is given by the following left-G_k-action on M via (4.11):

$$\begin{aligned}
G_k \times M &\longrightarrow M \\
(s\quad,\quad \varphi) &\mapsto s \circ \varphi.
\end{aligned} \quad (4.13)$$

Proof.   According to lemma 4.16, we have

$$Z^1_{cont}(S, S_n) = \mathrm{Hom}_{cont}(S, S_n) \cong \mathrm{Hom}_{cont}(S, \mathrm{Aut}(M)),$$

so continuous 1-cocycles of S in S_n correspond to continuous left-S-actions on M. Obviously two such actions are isomorphic if and only if the corresponding group homomorphisms $S \to S_n$ are conjugated, which according to lemma 4.16 is the case if and only if they are cohomologous as cocycles — this proves (4.11). (Note that even though the bijection $S_n \cong \mathrm{Aut}(M)$ is *not* canonical, the choice of another bijection leads to isomorphic left-S-actions, so that we indeed get a *canonical* isomorphism in (4.11).)

The rest follows immediately from proposition 4.3. $\qquad\square$

Lemma 4.17   *As in corollary 4.7, let k be a field with separable closure K and absolute Galois group $G := G_k$. Moreover, let $n \in \mathbb{N}_+$ be a positive natural number, and let A be a discrete G-group, so that according to (4.7)* *we get the following short exact sequence of exact G-groups:*

$$1 \longrightarrow A^n \xrightarrow{i_{A^n}} A \wr S_n \xrightarrow{p} S_n \longrightarrow 1.$$
$$\overset{i_{S_n}}{\longleftarrow}$$

Let $\gamma \in Z^1_{cont}(G, S_n)$ *be an arbitrary continuous 1-cocycle, and denote its image* $i_{S_n} \circ \gamma$ *in* $Z^1_{cont}(G, A \wr S_n)$ *by* c.

According to proposition 4.2(2), we then get the following two short exact sequences (of groups respectively pointed sets):

$$1 \longrightarrow H^0(G, A^r_c) \xrightarrow{H^0(i_{A^r})} H^0(G, (A \wr S_n)_c) \xrightarrow{H^0(p)} H^0(G, S_n) \longrightarrow 1$$
$$\overset{H^0(i_{S_n})}{\longleftarrow}$$

$$* \longrightarrow H^1_{cont}(G, A^r_c) \xrightarrow{H^1_{cont}(i_{A^r})} H^1_{cont}(G, (A \wr S_n)_c) \xrightarrow{H^1_{cont}(p)} H^1_{cont}(G, S_n) \longrightarrow *$$
$$\overset{H^1_{cont}(i_{S_n})}{\longleftarrow}$$

According to corollary 4.7, the cohomology class $[c]$ *corresponds to (the k-isomorphism class of) an fcs-algebra* L *of degree* n *over* k.

As in corollary 4.7, let M *denote the set* $Hom_{k\text{-}alg}(L, K)$ *with* n *elements. We want to fix a bijection* $\{1, 2, \ldots, n\} \xrightarrow{\sim} M$ *and identify the symmetric group* S_n *with* $Aut(M)$ *and* A^n *with* A^M *via this bijection. We then can write elements of* A^n_c *in the form* $(x_\varphi)_{\varphi \in M}$ *(with* $x_\varphi \in A$) *and elements of* $A \wr S_n$ *in the form* $(x_\varphi)_{\varphi \in M} \cdot \sigma$ *with* $x_\varphi \in A$ *and* $\sigma \in Aut(M)$ *as elements of* $A \wr Aut(M)$. *With these identifications, the action of* G *on the twists* A^n_c *and* $(S_n)_c$ *is given as follows:*

$$G \times A^n_c \longrightarrow A^n_c, \quad \left(s, (x_\varphi)_{\varphi \in M}\right) \mapsto ({}^s x_{s^{-1} \circ \varphi})_{\varphi \in M}$$
$$G \times (S_n)_c \longrightarrow (S_n)_c, (s, \sigma) \mapsto \left[M \to M, \ \varphi \mapsto s \circ \sigma(s^{-1} \circ \varphi)\right].$$
$$(4.14)$$

In future we always want to use these identifications.

Proof. This is obvious from corollary 4.7, equation (4.9) and definition 4.8! $\qquad \square$

Lemma 4.18 *As in corollary 4.7 and lemma 4.17, let* k *be a field with separable closure* K *and absolute Galois group* G. *Moreover, let* $n, m \in \mathbb{N}_+$

be positive natural numbers, let $n_1, \ldots, n_m \in \mathbb{N}_+$ be positive natural numbers with $\sum_{i=1}^{m} n_i = n$, and let $L_1, \ldots, L_m \subseteq K$ separable field extensions of k with $[L_i : k] = n_i$. Let L denote the product $\prod_{i=1}^{m} L_i$ which is an fcs-algebra of degree n over k.

According to corollary 4.7, the fcs-algebras L_i define cohomology classes $[c_i]$ with $c_i \in Z_{cont}^1(G, S_{n_i})$, and L defines a cohomology class $[c]$ with $c \in Z_{cont}^1(G, S_n)$.

Now let A be a discrete G-group. We then get the discrete G-groups $A_{c_i}^{n_i}$ and A_c^n defined in lemma 4.17. Put $M_i := Hom_k(L_i, K)$, $M := Hom_k(L, K)$, and let p_i denote the projection $L \twoheadrightarrow L_i$. Then the maps

$$M_i \longrightarrow M, \quad \varphi \mapsto \varphi \circ p_i$$

induce a bijection $\coprod_{i=1}^{m} M_i \xrightarrow{\sim} M$ and accordingly a natural group isomorphism

$$A_c^n \quad \xrightarrow{\sim} \quad \prod_{i=1}^{m} A_{c_i}^{n_i}$$
$$(x_\varphi)_{\varphi \in M} \quad \mapsto \quad [(x_{\varphi \circ p_i})_{\varphi \in M_i}]_{i=1,\ldots,m} \, .$$

This group isomorphism is an isomorphism of discrete G-groups.

Proof. We only have to show that the group isomorphism defined in the lemma is a morphism of (discrete) G-groups. For this it suffices to prove that for every i, the group homomorphism $A_c^n \twoheadrightarrow A_{c_i}^{n_i}$ (which we here want to denote by ψ_i) is a G-equivariant morphism.

So let $i \in \{1, \ldots, m\}$, $s \in G$ and $(x_\varphi)_{\varphi \in M} \in A_c^n$ be arbitrary elements. Then it follows:

$$\psi_i \left[{}^s(x_\varphi)_{\varphi \in M} \right] = \psi_i \left[({}^s x_{s^{-1}\varphi})_{\varphi \in M} \right] = ({}^s x_{s^{-1}\varphi p_i})_{\varphi \in M_i}$$
$$= {}^s[(x_{\varphi p_i})_{\varphi \in M_i}] = {}^s(\psi_i \left[(x_\varphi)_{\varphi \in M} \right]). \quad \square$$

So far, we have worked with one fixed topological group S and studied the functorial behaviour of the nonabelian cohomology of S with respect to morphisms between varying discrete S-groups and short exact sequences of discrete S-groups.

Now we briefly want to study what happens if we vary the topological group S itself; in particular, if $N \trianglelefteq S$ is a closed normal subgroup of S and

A is a discrete S-group, we want to understand the relation between the cohomologies $H^1_{\text{cont}}(S, A)$, $H^1_{\text{cont}}(N, A)$ and $H^1_{\text{cont}}(S/N, A^N)$. In the abelian case, this is achieved by the "restriction-inflation-sequence", and we will see that we get (the first terms of) that sequence in the nonabelian case, too.

Definition 4.10 Let R be another topological group, let $R \xrightarrow{\varphi} S$ be a continuous group homomorphism, and let A be a discrete S-group. Then the composition

$$R \xrightarrow{\varphi} S \longrightarrow \text{Aut}(A)$$

defines a continuous left-R-action on A — we want to call the resulting discrete R-group $\varphi^* A$. The identity morphism on A respectively the composition of cocycles with φ induces canonical morphisms (of groups respectively pointed sets)

$$H^0(S, A) \xrightarrow{\text{can}} H^0(R, \varphi^* A) \quad \text{and} \quad H^1_{\text{cont}}(S, A) \xrightarrow{\text{can}} H^1_{\text{cont}}(R, \varphi^* A)$$

$$a \quad\quad \mapsto \quad a \quad\quad\quad\quad [(a_s)] \quad\quad \mapsto \quad [(a_s) \circ \varphi] = [(a_{\varphi(r)})_r].$$

Now let B be a discrete R-group, and let $A \xrightarrow{f} B$ be a group homomorphism. We say that f is *compatible (with φ)* if $\varphi^* A \xrightarrow{f} B$ is a morphism of discrete R-groups, i.e. if

$$\forall r \in R: \ \forall a \in A: f\big[{}^{\varphi(r)}a\big] = {}^r\big[f(a)\big].$$

In that case, the compositions

$$H^0(S, A) \xrightarrow{\text{can}} H^0(R, \varphi^* A) \xrightarrow{H^0(f)} H^0(R, B) \quad \text{and}$$

$$H^1_{\text{cont}}(S, A) \xrightarrow{\text{can}} H^1_{\text{cont}}(R, \varphi^* A) \xrightarrow{H^1_{\text{cont}}(f)} H^1_{\text{cont}}(R, B)$$

define canonical morphisms $H^0(S, A) \to H^0(R, B)$ and $H^1_{\text{cont}}(S, A) \to H^1_{\text{cont}}(R, B)$ of groups respectively pointed sets.

Definition 4.11 Let A be a discrete S-group.

(1) Let $R \leq S$ be a closed subgroup of S, and denote the inclusion $R \hookrightarrow S$ (which is a continuous group homomorphism) by i. Then the identity on A, considered as a group homomorphism $i^* A \to A$, is compatible

with i, and we call the resulting canonical morphisms from definition 4.10 the *restriction* and denote it by res:

$$H^0(S, A) \xrightarrow{\text{res}} H^0(R, A) \quad \text{and} \quad H^1_{\text{cont}}(S, A) \xrightarrow{\text{res}} H^1_{\text{cont}}(R, A)$$

$$a \quad \mapsto \quad a \qquad\qquad [(a_s)] \quad \mapsto \quad [(a_s)|R] = [(a_r)].$$

(2) Let $N \trianglelefteq S$ be a closed *normal* subgroup of S, and denote the projection $S \twoheadrightarrow S/N$ (which is a continuous group homomorphism) by p. Then the group $A^N = H^0(N, A)$ becomes a discrete S/N-group by setting $^{\bar{s}}a := {}^s a$, and the inclusion $A^N \hookrightarrow A$, considered as a group homomorphism $p^* A^N \to A$, is compatible with p. The resulting morphisms from definition 4.10 are called *inflation* and denoted by inf:

$$H^0(S/N, A^N) \xrightarrow{\text{inf}} H^0(S, A) \quad \text{und} \quad H^1_{\text{cont}}(S/N, A^N) \xrightarrow{\text{inf}} H^1_{\text{cont}}(S, A)$$

$$a \quad \mapsto \quad a \qquad\qquad [(a_{\bar{s}})_{\bar{s}}] \quad \mapsto \quad [(a_{\bar{s}})_s].$$

Proposition 4.4   *Let A be a discrete S-group, and let $N \trianglelefteq S$ be a closed normal subgroup of S. Then the following sequence is exact in the category of pointed sets, and the inflation is injective:*

$$H^1_{cont}(S/N, A^N) \overset{\text{inf}}{\hookrightarrow} H^1_{cont}(S, A) \xrightarrow{\text{res}} H^1_{cont}(N, A).$$

Proof.

- *inflation injective:* Let $(a_{\bar{s}})$ and $(a'_{\bar{s}})$ be two continuous 1-cocycles of S/N in A^N whose cohomology classes are mapped to the same class under the inflation. Then there is a $b \in A$ with

$$\forall s \in S : \quad a'_{\bar{s}} = b^{-1} \cdot a_{\bar{s}} \cdot {}^s b. \tag{4.15}$$

We want to show that $b \in A^N$, because then (4.15) shows that $(a_{\bar{s}})$ and $(a'_{\bar{s}})$ are cohomologous. For that, let $t \in N$ be an arbitrary element. Then $\bar{t} = \bar{1}_S \in S/N$, and it follows:

$$1_A = a'_{\bar{1}_S} \overset{\text{Lemma 4.2}}{=} a'_{\bar{t}} \overset{(4.15)}{=} b^{-1} \cdot a_{\bar{t}} \cdot {}^t b$$

$$= b^{-1} \cdot a_{\bar{1}_S} \cdot {}^t b \overset{\text{Lemma 4.2}}{=} b^{-1} \cdot {}^t b \quad \Longrightarrow \quad {}^t b = b.$$

- *exact:* For an arbitrary class $[(a_{\bar{s}})]$ in $H^1_{\mathrm{cont}}(S/N, A^N)$ and an arbitrary $t \in N$, we have $a_{\bar{t}} = a_{\bar{1}_S} = 1_A$ because of Lemma 4.2, so $(\mathrm{res} \circ \mathrm{inf})[(a_{\bar{s}})]$ is the trivial class.

Conversely, let $[(a_s)] \in H^1_{\mathrm{cont}}(S, A)$ be a class whose restriction is the trivial class. Then there is a $b \in A$ with $b^{-1}a_t{}^t b = 1_A$ for all $t \in N$. Therefore, by replacing (a_s) with the cohomologous cocycle $(b^{-1}a_s{}^s b)$, we can assume that $a_t = 1_A$ for all $t \in N$ without loss of generality.

This gives us:

$$\forall s \in S \ \forall t \in N : a_{st} = a_s{}^s a_t = a_s{}^s 1_A = a_s. \tag{4.16}$$

Now let $s \in S$ and $t \in N$ be arbitrary elements. Because N is a normal subgroup of S, there is a $t' \in N$ with $ts = st'$, and we get:

$$a_s \stackrel{(4.16)}{=} a_{st'} = a_{ts} = a_t{}^t a_s = 1_A \cdot {}^t a_s = {}^t a_s.$$

So for all $s \in S$, the element a_s lies in A^N, and it follows from equation (4.16), that $(a_s)_{\bar{s}}$ is a well defined 1-cocycle of S/N in A^N. Then obviously this cocycle is a preimage of (a_s) under the inflation. □

Lemma 4.19 *Let A be a discrete S-group.*

(1) Let $N_1, N_2 \trianglelefteq S$ be to closed normal subgroups of S with $N_2 \subseteq N_1$. Then inflation defines a canonical map

$$H^1_{cont}(S/N_1, A^{N_1}) \xrightarrow{\ \inf\ } H^1_{cont}(S/N_2, A^{N_2}).$$

(2) Let N_1, N_2 and N_3 be closed normal subgroups of S with $N_3 \subseteq N_2 \subseteq N_1$. Then the following diagram commutes:

$$
\begin{array}{ccc}
H^1_{cont}(S/N_1, A^{N_1}) & \xrightarrow{\ \inf\ } & H^1_{cont}(S/N_2, A^{N_2}) \\
& \searrow{\scriptstyle \inf} & \big\downarrow{\scriptstyle \inf} \\
& & H^1_{cont}(S/N_3, A^{N_3})
\end{array}
$$

(3) Let $\mathcal{N} := \mathcal{N}_s$ be the set of closed normal subgroups of S of finite Index. Then $\mathcal{N}$ is a directed set with respect to the order $N_1 \leq N_2 :\Leftrightarrow N_2 \subseteq N_1$, and $\big(H^1_{cont}(S/N, A^N)\big)_{N \in \mathcal{N}}$ is an inductive system with respect to inflation.

Proof.   To see (1), just note that N_1/N_2 is a closed normal subgroup of S/N_2 and that we have

$$S/N_1 \cong (S/N_2)/(N_1/N_2) \quad \text{and} \quad A^{N_1} = (A^{N_2})^{N_1/N_2}.$$

With this we immediately get our desired map

$$H^1_{\text{cont}}(S/N_1, A^{N_1}) \cong H^1_{\text{cont}}\left[(S/N_2)/(N_1/N_2), (A^{N_2})^{N_1/N_2}\right]$$
$$\xrightarrow{\text{inf}} H^1_{\text{cont}}(S/N_2, A^{N_2}),$$

and it is obvious from this and from the definition of inflation that (2) holds.

To show (3), let N_1 and N_2 be two elements of $\mathcal{N}$. It suffices to show that $N_1 \cap N_2$ is in $\mathcal{N}$ as well, and because $N_1 \cap N_2$ is obviously a closed normal subgroup of S, we only have to see that $N_1 \cap N_2$ is of finite index in S. This is easy, since we have a canonical monomorphism $N_1/N_1 \cap N_2 \hookrightarrow S/N_2$ — which implies $(N_1 : N_1 \cap N_2) \le (S : N_2)$ — and since we know

$$(S/N_1 \cap N_2)/(N_1/N_1 \cap N_2) \cong S/N_1.$$

From these two facts we conclude

$$(S : N_1 \cap N_2) = (S : N_1) \cdot (N_1 : N_1 \cap N_2) \le (S : N_1) \cdot (S : N_2) \le \infty. \qquad \square$$

Proposition 4.5   *Let F be a functor from a small non-empty filtered category $\mathcal{C}$ to the category of pointed sets. Then the colimit $\varinjlim F$ exists in the category of pointed sets.*

Proof.   Let M be the colimit of F taken in the category of sets, i.e. $M = \left(\bigsqcup_{X \in \text{Ob}(\mathcal{C})} F(X)\right) / \sim$, where $x \in F(X)$ and $y \in F(Y)$ are equivalent if and only if there is an object $Z \in \text{Ob}(\mathcal{C})$ and morphisms $f : X \to Z$ and $g : Y \to Z$ such that $[F(f)](x) = [F(f)](y)$ in $F(Z)$.

By hypothesis, $\mathcal{C}$ is not empty, so there is an object X of $\mathcal{C}$, and we define the image of the special point x of $F(X)$ in M to be the special point of M. This is well defined: Let Y be another object of $\mathcal{C}$, and let y be the special point of $F(Y)$. Then because $\mathcal{C}$ is filtered, there is an object Z in $\mathcal{C}$ and morphisms $f : X \to Z$ and $g : Y \to Z$. Since f and g are morphisms of pointed sets, both x and y are mapped to the special point of $F(Z)$ under

f respectively g, so x and y define the same element of M.

Having defined the special point of M in this way, it is clear that the canonical maps $i_X : F(X) \to M$ for $X \in \mathrm{Ob}(\mathcal{C})$ become morphisms of pointed sets.

Now let N be a pointed set, and let $(f_X : F(X) \to N)_{X \in \mathrm{Ob}(\mathcal{C})}$ be a compatible family of morphisms of pointed sets. Then by the universal property of M in the category of sets, there is a unique map $f : M \to N$ satisfying $f_X = f \circ i_X$ for all objects X of $\mathcal{C}$. All that remains to be shown is that f is a morphism of pointed sets, but this is clear from the definition of the special point of M. $\qquad\square$

Corollary 4.8 *Let A be a discrete S-group. Because of lemma 4.19(3) and proposition 4.5, we can consider the colimit $\varinjlim_{N \in \mathcal{N}} H^1_{cont}(S/N, A^N)$.*

(1) Inflation induces a canonical injection of pointed sets

$$\varinjlim_{N \in \mathcal{N}} H^1_{cont}(S/N, A^N) \hookrightarrow H^1_{cont}(S, A).$$

(2) If S is a profinite group, then the injection from (1) is an isomorphism of pointed sets.

Proof.   For each $N \in \mathcal{N}$, inflation defines a morphism $H^1_{\mathrm{cont}}(S/N, A^N) \to H^1_{\mathrm{cont}}(S, A)$, and it is obvious from the definition of inflation that these morphisms are compatible and therefore define a morphism $\varinjlim H^1_{\mathrm{cont}}(S/N, A^N) \to H^1_{\mathrm{cont}}(S, A)$.

To prove injectivity, let x and y be two elements of $\varinjlim H^1_{\mathrm{cont}}(S/N, A^N)$ which are mapped to the same element in $H^1_{\mathrm{cont}}(S, A)$. Without loss of generality we can assume that both x and y have representatives in the same cohomology $H^1_{\mathrm{cont}}(S/N, A^N)$. But then they must be equal according to proposition 4.4.

Now let S be profinite. Then a normal subgroup N of S is in $\mathcal{N}$ if and only if it is open, and in addition to that we know that the elements of $\mathcal{N}$ form a fundamental system of open neighbourhoods of 1_S in S. Let x be an arbitrary element of $H^1_{\mathrm{cont}}(S, A)$, represented by a continuous 1-cocycle $(a_s)_s$, and let $U := \{s \in S | a_s = 1_A\}$. Because $(a_s)_s$ is continuous and $\{1_A\} \subseteq A$ is open, U must be open, and because $1_S \in U$ according to lemma 4.2, there is an $N \in \mathcal{N}$ with $N \subseteq U$. But then the restriction of x to $H^1_{\text{cont}}(N, A)$ is the trivial class, so proposition 4.4 shows that x lies in the image of $H^1_{\text{cont}}(S/N, A^N)$ under inflation. This shows that $\varinjlim H^1_{\text{cont}}(S/N, A^N) \to H^1_{\text{cont}}(S, A)$ is surjective and concludes the proof of the corollary. $\qquad\square$

Chapter 5

Weil cohomology theories and *l*-adic cohomology

In this chapter, we want to give a summary of those basic facts on *l*-adic cohomology that we will need in this book. The standard references on this topic are SGA 4, SGA 4 1/2 and SGA 5 ([GV72], [GVSD73], [GAD73], [GDB+77] and [GJIB77]), as well as the textbooks [Mil80], [FK88] and [KW01]; a short review can be found in [Kat91].

One crucial property of *l*-adic cohomology is its being a *Weil cohomology theory* if the base field is separably closed. We are going to explain what that means and how this implies important theorems, notably the *Lefschetz trace formula* in a purely formal way (for this compare [Kle91]).

Let k be a field with separable closure $\bar{k}$ and absolute Galois group $G_k :=$ Gal$(\bar{k}/k)$, and let l be a prime number which differs from the characteristic of k.

Definition 5.1 *(Étale cohomology)*
Let X be a scheme.

(1) The *(small) étale site* $X_{\text{ét}}$ has as underlying category $\mathbf{\acute{E}t}/X$, whose objects are the étale morphisms $U \to X$ and whose morphisms are X-morphisms. The coverings are the surjective families of étale morphisms $(U_i \to U)_{i \in I}$ in $\mathbf{\acute{E}t}/X$.

(2) An *(abelian) étale sheaf on X* is an abelian sheaf on $X_{\text{ét}}$; we denote the category of abelian étale sheaves on X by $\mathbf{S_{\text{ét}}}(X)$.

(3) Let A be an abelian group. Then the étale sheaf associated to the constant presheaf $U \mapsto A$ is denoted by A_X.

(4) Let $\mathcal{F}$ be an abelian étale sheaf on X, and let $i \in \mathbb{N}_0$ be a natural

number. Then the *i-th étale cohomology of X with coefficients in $\mathcal{F}$* is

$$\mathrm{H}^i_{\text{ét}}(X, \mathcal{F}) := R^i\Gamma(X, \_)(\mathcal{F}).$$

(5) Let $f : X \to Y$ be a morphism of schemes. We get an induced morphism of sites $Y_{\text{ét}} \to X_{\text{ét}}$ be sending $U \to Y$ to $U \times_Y X \to X$ and hence an induced pair of adjoint functors $f_* : \mathbf{S}_{\text{ét}}(X) \to \mathbf{S}_{\text{ét}}(Y)$ and $f^* : \mathbf{S}_{\text{ét}}(Y) \to \mathbf{S}_{\text{ét}}(X)$.

Let $f : X \to Y$ be a morphism of schemes, and let $\mathcal{G}$ be an abelian étale sheaf on Y. Then we have the Grothendieck spectral sequence for the composition of functors $\mathbf{S}_{\text{ét}}(X) \xrightarrow{f_*} \mathbf{S}_{\text{ét}}(Y) \xrightarrow{\Gamma(Y,\_)} \underline{\mathrm{Ab}}$, applied to the sheaf $f^*\mathcal{G}$ on X,

$$\mathrm{H}^p_{\text{ét}}(Y, R^q f_* f^*\mathcal{G}) \Longrightarrow R^{p+q} \underbrace{(\Gamma(Y, \_) \circ f_*)}_{=\Gamma(X,\_)} f^*\mathcal{G} = \mathrm{H}^{p+q}_{\text{ét}}(X, f^*\mathcal{G}),$$

and thus for $i \geq 0$ an edge morphism

$$\mathrm{H}^i_{\text{ét}}(Y, f_* f^*\mathcal{G}) \longrightarrow \mathrm{H}^i_{\text{ét}}(X, f^*\mathcal{G}). \tag{5.1}$$

On the other hand, the unit of the adjunction $f^* \dashv f_*$ by functorality induces a morphism

$$\mathrm{H}^i_{\text{ét}}(Y, \mathcal{G}) \longrightarrow \mathrm{H}^i_{\text{ét}}(Y, f_* f^*\mathcal{G}).$$

Composing this with the edge morphism (5.1), we get a morphism

$$\mathrm{H}^i_{\text{ét}}(Y, \mathcal{G}) \longrightarrow \mathrm{H}^i_{\text{ét}}(X, f^*\mathcal{G}). \tag{5.2}$$

Now let A be a finite abelian group, and let $\mathcal{G} := A_Y$. Then f^*A_Y is canonically isomorphic to A_X, so that from (5.2) we get

$$f^* : \mathrm{H}^i_{\text{ét}}(Y, A_Y) \longrightarrow \mathrm{H}^i_{\text{ét}}(X, A_X). \tag{5.3}$$

This shows that $X \mapsto \mathrm{H}^i_{\text{ét}}(X, A_X)$ defines a contravariant functor from the category of schemes to the category of abelian groups.

For a k-variety X, put $\bar{X} := X \times_k \bar{k}$. The canonical left-$G_k$-action on $\bar{k}$ induces a right-G_k-action on $\bar{X}$ and thus a left-G_k-action on $\mathrm{H}^i_{\text{ét}}(\bar{X}, A_{\bar{X}})$. It is easy to see that this G_k-action is continuous with respect to the discrete topology on $\mathrm{H}^i_{\text{ét}}(\bar{X}, A_{\bar{X}})$.

If $f : X \to Y$ is a morphism of k-varieties, and if $s \in G_k$, then the diagram

$$\begin{array}{ccc} \bar{X} & \xrightarrow{\bar{f}} & \bar{Y} \\ {\scriptstyle s^*}\downarrow & & \downarrow{\scriptstyle s^*} \\ \bar{X} & \xrightarrow{\bar{f}} & \bar{Y} \end{array}$$

obviously commutes, so that $f^* : \mathrm{H}^i_{\text{ét}}(\bar{Y}, A_{\bar{Y}}) \to \mathrm{H}^i_{\text{ét}}(\bar{X}, A_{\bar{X}})$ is G_k-equivariant. This shows that $X \mapsto \mathrm{H}^i_{\text{ét}}(\bar{X}, A_{\bar{X}})$ defines a contravariant functor from the category of k-varieties to the category of discrete G_k-modules.

In the special case where $l^n \cdot A = 0$ for an $n \in \mathbb{N}_+$, by functorality we have $l^n \cdot \mathrm{H}^i_{\text{ét}}(\bar{X}, A_{\bar{X}}) = 0$, so that $\mathrm{H}^i_{\text{ét}}(\bar{X}, A_{\bar{X}})$ is a $\mathbb{Z}/l^n$-module.

Now we can define l-adic cohomology:

Definition 5.2 *(L-adic cohomology)*
Let X be a k-variety, and let $i \in \mathbb{N}_0$ be a natural number. For every $n \in \mathbb{N}_+$, the natural group homomorphism $\mathbb{Z}/l^{n+1} \to \mathbb{Z}/l^n$ induces a morphism $\mathrm{H}^i_{\text{ét}}(\bar{X}, (\mathbb{Z}/l^{n+1})_{\bar{X}}) \to \mathrm{H}^i_{\text{ét}}(\bar{X}, (\mathbb{Z}/l)_{\bar{X}})$, and we define

$$\mathrm{H}^i_{\text{ét}}(\bar{X}, \mathbb{Z}_l) := \varprojlim_{n \in \mathbb{N}_+} \mathrm{H}^i_{\text{ét}}(\bar{X}, (\mathbb{Z}/l)_{\bar{X}}),$$

which is a $\mathbb{Z}_l$-module with continues G_k-action (with respect to the l-adic topology). We define the *i-th l-adic cohomology of X* as

$$\mathrm{H}^i_{\text{ét}}(\bar{X}, \mathbb{Q}_l) := \mathrm{H}^i_{\text{ét}}(\bar{X}, \mathbb{Z}_l) \otimes_{\mathbb{Z}_l} \mathbb{Q}_l.$$

Thus l-adic cohomology is a contravariant functor from the category of k-varieties to the category of $\mathbb{Q}_l$-vector spaces with l-adically continuous G_k-action.

Readers not familiar with the formalism of Grothendieck topologies do not have to worry: In this book, it will not be important for us *how* the functor of l-adic cohomology is constructed, we only need its most important properties.

Definition 5.3 *(Arithmetic and geometric Frobenius)*
Consider the special case where $k = \mathbb{F}_q$ is a finite field, let X be a *projective* k-variety in $\mathbb{P}_k^n$, and let $i \in \mathbb{N}_0$ be a natural number.

(1) Let f be the Frobenius $x \mapsto x^q$ in G_k. Then, because of the G_k-action on l-adic cohomology, we get an induced automorphism f_X^* on $\mathrm{H}_{\text{ét}}^i(\bar{X}, \mathbb{Q}_l)$. We call both f and f_X the *arithmetic Frobenius*.

(2) Let $F_X : \bar{X} \to \bar{X}$ be the $\bar{k}$-morphism which on $\bar{k}$-rational points is given by $[x_0 : \ldots : x_n] \mapsto [x_0^q : \ldots : x_n^q]$. Then F_X and the induced endomorphism F_X^* of $\mathrm{H}_{\text{ét}}^i(\bar{X}, \mathbb{Q}_l)$ are called the *geometric Frobenius*.

Proposition 5.1 *In the situation of definition 5.3, the geometric Frobenius F_X^* is actually an automorphism whose inverse is the arithmetic Frobenius f_X^*.*

Remark 5.1 *Proposition 5.1 says that $(f^{-1})^* = F_X^*$ on $\mathrm{H}_{\text{ét}}^i(\bar{X}, \mathbb{Q}_l)$. For this reason, we sometimes also call f^{-1} the* geometric Frobenius *of G_k.*

If k is separably closed, then the G_k-action on l-adic cohomology is trivial, and we just get a functor into the category of $\mathbb{Q}_l$-vector spaces.

If we restrict this functor to the category $\mathcal{C}_{\bar{k}}$ of smooth, projective, irreducible varieties over $\bar{k}$, then we get a *Weil cohomology theory* — we will briefly recall what that means (see [Kle91]):

Definition 5.4 *(Weil cohomology theory)*
Let L be a field of characteristic zero, and let

$$X \mapsto H^*(X), \quad [X \xrightarrow{f} Y] \mapsto [H^*(Y) \xrightarrow{f^*} H^*(X)]$$

be a contravariant functor from $\mathcal{C}_{\bar{k}}$ to the category of graded anticommutative L-algebras. This functor is called a *Weil cohomology theory (on $\mathbb{C}_{\bar{k}}$, with coefficients in L* if it has the following properties:

(1) *(Finiteness):* Each $H^i(X)$ has finite dimension, and vanishes unless $0 \leq i \leq 2 \dim X$.

(2) *(Poincaré duality):* For each X of dimension r, there is a functorial "orientation" isomorphism $H^{2r}(X) \xrightarrow{\sim} L$, $u \mapsto \langle u \rangle$, which on

$L \cong H^0(\operatorname{Spec} \bar{k})$ is the identity, that induces a nondegenerate L-bilinear pairing

$$H^i(X) \times H^{2r-i}(X) \longrightarrow L, \quad (x,y) \mapsto \langle x \cdot y \rangle$$

for $0 \le i \le 2r$. For convenience, given a Y of dimension s and a morphism $f : X \to Y$, let

$$f_* : H^i(X) \to H^{2s-2r+i}(Y)$$

denote the dual of $f^* : H^{2r-i}(Y) \to H^{2r-i}(X)$ with respect to this pairing, i.e. the L-linear map making the following diagram commutative:

$$
\begin{array}{ccccc}
H^{2r-i}(X) & \times & H^i(X) & \xrightarrow{\langle\,,\,\rangle} & L \\
{\scriptstyle f^*}\big\uparrow & & \big\downarrow{\scriptstyle f_*} & & \big\| \\
H^{2r-i}(Y) & \times & H^{2s-2r+i}(Y) & \xrightarrow[\langle\,,\,\rangle]{} & L
\end{array}
\qquad (5.4)
$$

(3) *(Künneth formula):* For each X and Y, the projections $p_X : X \times_{\bar{k}} Y \to X$ and $p_Y : X \times_{\bar{k}} Y \to Y$ induce an isomorphism

$$H^*(X) \otimes_L H^*(Y) \xrightarrow[\sim]{\; x \otimes y \mapsto p_X^* x \cdot p_Y^* y \;} H^*(X \times_{\bar{k}} Y).$$

(4) *(Cycle map):* For each X, let $\operatorname{CH}^i(X)$ denote i-*th Chow group of* X, i.e. the group of algebraic cycles of codimension i on X modulo rational equivalence. Then there is a group homomorphism

$$\operatorname{cl}_X : \operatorname{CH}^i(X) \to H^{2i}(X),$$

called the *cycle map*, satisfying:

(a) *(functorality):* For each map $f : X \to Y$ with $\dim X = r$ and $\dim Y = s$, the following diagrams commute:

$$
\begin{array}{ccc}
\operatorname{CH}^i(Y) & \xrightarrow{f^*} & \operatorname{CH}^i(X) \\
{\scriptstyle \operatorname{cl}_Y}\big\downarrow & & \big\downarrow{\scriptstyle \operatorname{cl}_X} \\
H^{2i}(Y) & \xrightarrow[f^*]{} & H^{2i}(X)
\end{array}
\qquad
\begin{array}{ccc}
\operatorname{CH}^i(X) & \xrightarrow{f_*} & \operatorname{CH}^{s-r+i}(Y) \\
{\scriptstyle \operatorname{cl}_X}\big\downarrow & & \big\downarrow{\scriptstyle \operatorname{cl}_Y} \\
H^{2i}(X) & \xrightarrow[f_*]{} & H^{2(s-r+i)}(Y)
\end{array}
$$

(b) *(Multiplicativity):* For each X and Y, the following diagram commutes:

$$\begin{array}{ccc}
\mathrm{CH}^i(X) \otimes_{\mathbb{Z}} \mathrm{CH}^j(Y) & \xrightarrow{\;\alpha \otimes \beta \mapsto \alpha \times \beta\;} & \mathrm{CH}^{i+j}(X \times_{\bar{k}} Y) \\
{\scriptstyle \mathrm{cl}_X \otimes \mathrm{cl}_Y}\Big\downarrow & & \Big\downarrow{\scriptstyle \mathrm{cl}_{(X \times_{\bar{k}} Y)}} \\
H^{2i}(X) \otimes_L H^{2j}(Y) & \xhookrightarrow[\;x \otimes y \mapsto p_X^* x \cdot p_Y^* y\;]{} & H^{2(i+j)}(X \times_{\bar{k}} Y)
\end{array}$$

(c) *(Calibration):* If $P = \mathrm{Spec}\,(\bar{k})$ is a point, then $\mathrm{cl}_P : \mathrm{CH}^0(P) \to H^0(P)$ is the canonical inclusion of $\mathbb{Z}$ into L.

(5) *(Weak Lefschetz theorem):* Let $h : W \hookrightarrow X$ be the inclusion of a smooth hyperplane section, and set $r := \dim X$. Then $h^* : H^i(X) \to H^i(W)$ is an isomorphism for $i \leq r - 2$ and injective for $i = r - 1$.

(6) *(Strong Lefschetz theorem):* Let W be a smooth hyperplane section of X, set $r := \dim X$, and define the *Lefschetz operator*

$$\mathbf{L} : H^i(X) \to H^{i+2}(X), \quad x \mapsto x \cdot \mathrm{cl}_X(W).$$

Then, for $i \leq r$, the $(r-i)$-th iterate of $\mathbf{L}$ is an isomorphism

$$\mathbf{L}^{r-i} : H^i(X) \xrightarrow{\sim} H^{2r-i}(X).$$

Theorem 5.1   *The contravariant functor* $X \;\mapsto\; H^*_{\acute{e}t}(\bar{X}, \mathbb{Q}_l) \;:=\; \bigoplus_{i=0}^{\infty} H^i_{\acute{e}t}(\bar{X}, \mathbb{Q}_l)$ *induces a Weil cohomology theory on* $\mathcal{C}_{\bar{k}}$ *with coefficients in the field* $\mathbb{Q}_l$.

(Note that we have neither explained the multiplication on $H^*_{\acute{e}t}(\bar{X}, \mathbb{Q}_l)$ nor the cycle map).

    In the next part of this chapter, we want to demonstrate how to work with the axioms of a Weil cohomology theory by proving some simple facts and the *Lefschetz trace formula* which will be of great importance for us later.

    For this, we fix a Weil cohomology theory $X \mapsto H^*(X)$ on $\mathcal{C}_{\bar{k}}$ with coefficients in a field L. From theorem 5.1 we know that everything we prove in this context will in particular be true for l-adic cohomology.

Lemma 5.1 (Projection formula)
Let $f : X \to Y$ be a morphism in $\mathcal{C}_{\bar{k}}$. Then the following diagram commutes:

$$
\begin{array}{ccc}
H^*(X) \times H^*(X) & \xrightarrow{\ (x_1,x_2)\mapsto x_1 \cdot x_2\ } & H^*(X) \\
{\scriptstyle f^*}\uparrow \quad\ \ \downarrow{\scriptstyle f_*} & & \downarrow{\scriptstyle f_*} \\
H^*(Y) \times H^*(Y) & \xrightarrow[\ (y_1,y_2)\mapsto y_1 \cdot y_2\ ]{} & H^*(Y)
\end{array}
$$

This means that for $x \in H^*(X)$ and $y \in H^*(Y)$, we have

$$f_*(f^*y \cdot x) = y \cdot f_*x. \tag{5.5}$$

Proof. Let $\dim(X) = r$ and $\dim(Y) = s$. Without loss of generality, we can assume that $x \in H^i(X)$ and $y \in H^j(Y)$. Then $f^*y \cdot x \in H^{i+j}(X)$, so by (5.4), we have to show that for all $z \in H^{2r-i-j}(Y)$, the equation

$$\langle f^*z \cdot (f^*y \cdot x) \rangle = \langle z \cdot (y \cdot f_*x) \rangle$$

holds. But this is obvious, since

$$\langle f^*z \cdot (f^*y \cdot x) \rangle = \langle f^*(z \cdot y) \cdot x \rangle \overset{(5.4)}{=} \langle (z \cdot y) \cdot f_*x \rangle = \langle z \cdot (y \cdot f_*x) \rangle. \qquad \square$$

Lemma 5.2 *Let X be an object of $\mathcal{C}_{\bar{k}}$ of dimension r with structure morphism $\pi_X : X \to Spec\,\bar{k}$, and let $x \in H^{2r}(X)$. Then*

$$\langle x \rangle = (\pi_X)_*x \in L.$$

Proof.

$$\langle x \rangle = \langle 1 \cdot x \rangle = \langle \pi_X^* 1 \cdot x \rangle \overset{(5.4)}{=} \langle 1 \cdot (\pi_X)_*x \rangle = \langle (\pi_X)_*x \rangle = (\pi_X)_*x \in L. \qquad \square$$

Lemma 5.3 *Let X and Y be objects of $\mathcal{C}_{\bar{k}}$ of dimension r respectively s. For $d \in \{0, \ldots, 2r + 2s\}$, consider the canonical isomorphism ι_d of L-vector* *spaces given by*

$$H^d(Y \times_k X) \xrightarrow{K\ddot{u}nneth} \bigoplus_i H^{d-i}(Y) \otimes_L H^i(X)$$

$$\xrightarrow{Poincar\acute{e}} \bigoplus_i H^{(2s-d)+i}(Y)^\vee \otimes_L H^i(X)$$

$$\xrightarrow{can} \mathrm{Hom}_L\left[H^*(Y), H^{*+(d-2s)}(X)\right].$$

Then

(1) For $f \in H^d(Y \times_k X)$ and $y \in H^*(Y)$, we have

$$\left[\iota_d f\right](y) = (p_X)_*\left[p_Y^* y \cdot f\right].$$

(2) The diagram

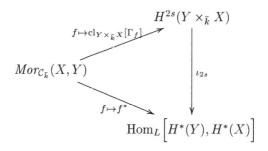

commutes. (For a morphism $f : X \to Y$, Γ_f denotes the graph of f, i.e. the image of the closed immersion $X \xrightarrow{(f,1_X)} Y \times_{\bar{k}} X$.)

Proof.   The map ι_d is obviously additive, so to check (1), without loss of generality we can assume that $f = p_Y^* z \cdot p_X^* x$ for $x \in H^i(X)$ and $z \in H^{d-i}(Y)$ and that $y \in H^{i-(d-2s)}(Y)$. Then

$$\left[\iota_d f\right](y) = \langle y \cdot z \rangle x \stackrel{\mathrm{lemma}\ 5.2}{=} \pi_X^* (\pi_Y)_*(y \cdot z) \cdot x$$

and

$$(p_X)_*\left[p_Y^* y \cdot f\right] = (p_X)_*\left[p_Y^* y \cdot p_Y^* z \cdot p_X^* x\right]$$

$$\stackrel{y \cdot z \in H^{2s}(Y)}{=} (p_X)_*\left[p_X^* x \cdot p_Y^*(y \cdot z)\right] \stackrel{(5.5)}{=} x \cdot (p_X)_* p_Y^*(y \cdot z),$$

so (1) would follow if we knew

$$\pi_X^*(\pi_Y)_* = (p_X)_* p_Y^* : H^{2s}(Y) \to H^0(X).$$

By L-linearity and calibration, it is enough to check that on $\mathrm{cl}_Y[P] \in H^{2s}(Y)$ for a closed point $P \in Y$. But then

$$(p_X)_* p_Y^* \mathrm{cl}_Y[P] = \mathrm{cl}_X\Big[(p_X)_* p_Y^*[P]\Big] = \mathrm{cl}_X\Big[(p_X)_*[P \times X])\Big] = \mathrm{cl}_X[X] = 1$$

and

$$\pi_X^*(\pi_Y)_* \mathrm{cl}_Y[P] = \mathrm{cl}_X\Big[\pi_X^*(\pi_Y)_*[P]\Big] = \mathrm{cl}_X\Big[\pi_X^*[\mathrm{Spec}\,\bar{k}]\Big] = \mathrm{cl}_X[X] = 1.$$

To prove (2), by (1) we have to show that for any $y \in H^*(Y)$, the following holds:

$$f^* y = (p_X)_* \Big[p_Y^* y \cdot \mathrm{cl}_{Y \times_{\bar{k}} X}[\Gamma_f]\Big].$$

Using $[\Gamma_f] = (f, 1_X)_*[X]$, we calculate:

$$(p_X)_* \underbrace{\Big[p_Y^* y \cdot \mathrm{cl}_{Y \times_{\bar{k}} X}[\Gamma_f]\Big]}_{\overset{\text{lemma } 5.1}{=} (f, 1_X)_* \Big[(f, 1_X)^* p_Y^* y \cdot \mathrm{cl}_X[X]\Big]}$$

$$= [p_X \circ (f, 1_X)]_* \Big[[p_Y \circ (f, 1_X)]^* y \cdot 1\Big] = (1_X)_* f^* y = f^* y.$$

$\square$

Definition 5.5 Let X be an object of $\mathcal{C}_{\bar{k}}$ of dimension r. Then by the Künneth isomorphism, the cycle class of the diagonal $\Delta : X \hookrightarrow X \times_{\bar{k}} X$ decomposes into a sum $\sum_{i=0}^{2r} \pi_X^{(i)}$ where π^i comes from $H^{2r-i}(X) \otimes_L H^i(X)$. These classes $\pi_X^{(0)}, \ldots, \pi_X^{(2r)}$ in $H^{2r}(X \times_{\bar{k}} X)$ are called the *Künneth components* of X.

Proposition 5.2 (Trace formula) *Let X be an object of $\mathcal{C}_{\bar{k}}$ of dimension r, and let u be an element of $H^{2r}(X \times_{\bar{k}} X)$. Then for all $d \in \{0, \ldots, 2r\}$, we have the following trace formula:*

$$Tr\left(\iota_{2r} u \middle| H^d(X)\right) = (-1)^d \cdot \left\langle u \cdot \pi_X^{(d)} \right\rangle. \tag{5.6}$$

Proof. For $i \in \{0, \ldots, 2d\}$, let $\{e_j^{(i)}\}$ be a basis of $H^i(X)$ with $e^{(2r)} := e_1^{(2r)}$ chosen such that $\langle e^{(2r)} \rangle = 1$, and let $\{f_j^{(2r-i)}\}$ be the dual basis of $H^{2r-i}(X)$, i.e. the one satisfying $\langle e_j^{(i)} \cdot f_k^{(2r-i)} \rangle = \delta_{jk}$ (and hence $e_j^{(i)} \cdot f_j^{(2r-i)} = e^{(2r)}$). Denote the projections $X \times_{\bar{k}} X \to X$ by p_1 and p_2 and put $u^* := \iota_{2r} u \in \operatorname{End}_L(H^*(X))$. Then by lemma 5.3(1),

$$u = \sum_{i=0}^{2r} \sum_j p_1^* f_j^{(2r-i)} \cdot p_2^* u^* e_j^{(i)}$$

and

$$\pi_X^{(d)} = \sum_j p_1^* f_j^{(2r-d)} \cdot p_2^* e_j^{(d)} = (-1)^d \cdot \sum_j p_1^* e_j^{(2r-d)} \cdot p_2^* f_j^{(d)}$$

Thus

$$u \cdot \pi_X^{(d)} = (-1)^d \sum_{i=0}^{2r} \sum_{j,k} p_1^* f_j^{(2r-i)} \cdot p_2^* u^* e_j^{(i)} \cdot p_1^* e_k^{(2r-d)} \cdot p_2^* f_k^{(d)}$$

$$= (-1)^{d+i(2r-d)} \sum_{i=0}^{2r} \sum_{j,k} p_1^* \underbrace{\left[f_j^{(2r-i)} \cdot e_k^{(2r-d)} \right]}_{\in H^{2r+(2r-i-d)}(X)} \cdot p_2^* \underbrace{\left[u^* e_j^{(i)} \cdot f_k^{(d)} \right]}_{\in H^{2r-(2r-i-d)}(X)}$$

$$= \underbrace{(-1)^{d+(2r-d)(2r-d)}}_{=1} \sum_{j,k} p_1^* \left[f_j^{(d)} \cdot e_k^{(2r-d)} \right] \cdot p_2^* \underbrace{\left[u^* e_j^{(2r-d)} \cdot f_k^{(d)} \right]}_{=(-1)^d \delta_{jk} \cdot e^{(2r)}}$$

$$= (-1)^d \sum_j p_1^* e^{(2r)} \cdot p_2^* \left[u^* e_j^{(2r-d)} \cdot f_j^{(d)} \right] \quad (5.7)$$

For all j, let $u^* e_j^{(2r-d)} = \sum_k a_{kj} e_k^{(2r-d)}$ with $a_{kj)}$ in L. Then we get

$$u \cdot \pi_X^{(d)} \overset{(5.7)}{=} (-1)^d \sum_j p_1^* e^{(2r)} \cdot p_2^* \left[\left(\sum_k a_{kj} e_k^{(2r-d)} \right) \cdot f_j^{(d)} \right]$$

$$= (-1)^d \sum_j p_1^* e^{(2r)} \cdot p_2^* \left[a_{jj} e^{(2r)} \right]$$

$$= (-1) \underbrace{\left(\sum_j a_{jj} \right)}_{=\operatorname{Tr}\left[u^* \big| H^i(X) \right]} \cdot p_1^* e^{(2r)} \cdot p_2^* e^{(2r)},$$

so

$$\left\langle u \cdot \pi_X^{(d)} \right\rangle = (-1)^d \cdot \mathrm{Tr}\left[u^* \middle| H^i(X)\right] \cdot \underbrace{\left\langle p_1^* e^{(2r)} \cdot p_2^* e^{(2r)} \right\rangle}_{=1}.$$

$\square$

As a corollary, we are now going to prove one of the most important theorems on *l*-adic cohomology, the *Lefschetz trace formula*, which will provide a link between *l*-adic cohomology and the zeta function of smooth projective varieties over finite fields.

As before, we will work in the more general context of our fixed arbitrary Weil cohomology theory.

Theorem 5.2   (Lefschetz trace formula)

Let $X \mapsto H^*(X)$ be a Weil cohomology theory on $C_{\bar{k}}$ with coefficients in a field L (for example, we can take l-adic cohomology $X \mapsto H_{\acute{e}t}^*(X, \mathbb{Q}_l)$), let X be an object of $C_{\bar{k}}$ of dimension r, and let $f : X \to X$ be a $\bar{k}$-morphism. Denote the graphs of f and 1_X, considered as r-cycles on $X \times_{\bar{k}} X$, by Γ_f respectively Δ. Then we have the following formula for the intersection number $(\Gamma_f \cdot \Delta)$:

$$(\Gamma_f \cdot \Delta) = \sum_{i=0}^{2r} (-1)^i \cdot Tr\left[f^* \middle| H^i(X)\right]. \tag{5.8}$$

If f only has isolated fixed points (which means that Γ_f meets Δ transversally), then $(\Gamma_f \cdot \Delta)$ is just the number of fixed points of f. For this reason, formula (5.8) is also called the "Lefschetz fixed point formula".

Proof.   We calculate

$$(\Gamma_f \cdot \Delta) = \left\langle \mathrm{cl}_{X \times_{\bar{k}} X}[\Gamma_f] \cdot \mathrm{cl}_{X \times_{\bar{k}} X}[\Delta] \right\rangle = \sum_{i=0}^{2r} \left\langle \mathrm{cl}_{X \times_{\bar{k}} X}[\Gamma_f] \cdot \pi_X^i \right\rangle$$

$$\stackrel{\text{proposition 5.2}}{=} \sum_{i=0}^{2r} (-1)^i \mathrm{Tr}\left[\iota_{2r}\left(\mathrm{cl}_{X \times_{\bar{k}} X}[\Gamma_f]\right) \middle| H^i(X)\right].$$

From this the claim follows, since $\iota_{2r}\left(\mathrm{cl}_{X \times_{\bar{k}} X}[\Gamma_f]\right) = f^*$ by lemma 5.3(2) $\square$

From now on, let $k = \mathbb{F}_q$ be a finite field, and let X be a smooth, projective k-variety of dimension r in $\mathbb{P}_k^n$.

If $P = [x_0 : \ldots : x_n]$ is a $\bar{k}$-rational point of X, then P is already $\mathbb{F}_{q^j}$-rational for a $j \in \mathbb{N}_+$ if and only if $[x_0 : \ldots : x_n] = [x_0^{(q^j)} : \ldots : x_n^{(q^j)}]$, which means that $\nu_X^{(j)} = \#X(\mathbb{F}_{q^j})$ is the number of fixed points of F_X^j, the j-th iterate of the geometric Frobenius on X.

It is easy to see that all fixed points of F_X^j are isolated, so that their number can be computed with (5.8), giving us the following formula for $\nu_X^{(j)}$:

$$\boxed{\nu_X^{(j)} = \sum_{i=0}^{2r} (-1)^i \cdot \mathrm{Tr}\left[(F_X^j)^* \,\middle|\, \mathrm{H}_{\text{ét}}^i(\bar{X}, \mathbb{Q}_l) \right].} \qquad (5.9)$$

Lemma 5.4   *Let L be a field of characteristic zero, let V be a finite-dimensional L-vector space, and let A be an L-linear endomorphism of V. Then*

$$\exp\left(\sum_{j=1}^{\infty} \frac{\mathrm{Tr}[A^j]}{j} T^j \right) = \frac{1}{\det(1_V - AT)} \in L[[T]].$$

Proof.   The lemma and the idea for its proof are from [Mil80, Lemma 2.7.,p186]. Let $\bar{L}$ be an algebraic closure of L. Obviously, the lemma is true for L, V and A if and only if it is true for $\bar{L}$, $V \otimes_L \bar{L}$ and $A \otimes 1_{\bar{L}}$, so without loss of generality, we can assume that L is algebraically closed, so that there is a basis of V with respect to which A is given by an upper triangular matrix with elements $\lambda_1, \ldots, \lambda_n$ on its diagonal. Using the formula

$$\log\left(\frac{1}{1-T} \right) = \sum_{j=1}^{\infty} \frac{T^j}{j}, \qquad (5.10)$$

we get

$$\log\left(\frac{1}{\det(1_V - AT)}\right) = \log\left(\frac{1}{\prod_{i=1}^{n}(1 - \lambda_i T)}\right) = \sum_{i=1}^{n}\log\left(\frac{1}{1 - \lambda_i T}\right)$$

$$\overset{(5.10)}{=} \sum_{i=1}^{n}\sum_{j=1}^{\infty}\frac{\lambda_i^j}{j}T^j = \sum_{j=1}^{\infty}\frac{\sum_{i=1}^{n}\lambda_i^n}{j}T^j = \sum_{j=1}^{\infty}\frac{\text{Tr}\left[A^j\right]}{j}T^j.$$

Taking the exponential of both sides completes the proof of the lemma. $\square$

Using lemma 5.4 and formula (5.9), we can now express the zeta function of X in terms of the characteristic polynomials of the geometric Frobenius F_X^*:

Corollary 5.1   *For $d \in \{0, \ldots, 2r\}$, put*

$$P_d := P_d^X := \det\left(1 - (F_X^d)^* T \,\middle|\, H_{\text{ét}}^d(\bar{X}, \mathbb{Q}_l)\right) \in 1 + T\mathbb{Q}_l[T] \subset \mathbb{Q}_l[T].$$

Then

$$\zeta(X, T) = \frac{P_1(T) \cdot P_3(T) \cdot \ldots \cdot P_{2r-1}(T)}{P_0(T) \cdot P_2(T) \cdot \ldots \cdot P_{2r}(T)}. \tag{5.11}$$

Proof.   For $j \in \mathbb{N}_+$ and $d \in \{0, \ldots, 2r\}$ put

$$t_d^{(j)} := \text{Tr}\left[(F_X^j)^* \,\middle|\, H_{\text{ét}}^d(\bar{X}, \mathbb{Q}_l)\right].$$

Then

$$\zeta(X, T) = \exp\left(\sum_{j=1}^{\infty}\frac{\nu_X^{(j)}}{j}T^j\right) \overset{(5.9)}{=} \exp\left(\sum_{j=1}^{\infty}\frac{\sum_{d=0}^{2r}(-1)^d t_d^{(j)}}{j}T^j\right)$$

$$= \prod_{d=0}^{2r}\left[\exp\left(\sum_{j=1}^{\infty}\frac{t_d^j}{j}T^j\right)\right]^{[(-1)^d]} \overset{\text{lemma 5.4}}{=} \prod_{d=0}^{2r}\left[\frac{1}{P_d(T)}\right]^{[(-1)^d]}$$

$$= \prod_{d=0}^{2r}P_d(T)^{[(-1)^{d+1}]} = \frac{P_1(T) \cdot P_3(T) \cdot \ldots \cdot P_{2r-1}(T)}{P_0(T) \cdot P_2(T) \cdot \ldots \cdot P_{2r}(T)}.$$

$\square$

Theorem 5.3   *For all* $d \in \{0, \ldots, 2r\}$, *the characteristic polynomials* $P_d(T)$ *are elements of* $\mathbb{Z}[T]$, *and if we write*

$$P_d(T) = \prod_{i=1}^{\beta_d} (1 - \alpha_{d,i} T),$$

then the $\alpha_{d,i}$ *are algebraic integers of absolute value* $q^{\frac{d}{2}}$ *with respect to every embedding* $\bar{\mathbb{Q}} \hookrightarrow \mathbb{C}$.

Proof.   This was proved by Deligne in [Wei49].   □

Using corollary 5.1 and theorem 5.3, it is easy to prove the first three Weil conjecture (see remark 2.1): *Rationality* follows immediately from (5.11) and the fact that the polynomials $P_d(T)$ have integer coefficients. The *functional equation* is an easy consequence of Poincaré duality, the *Riemann hypothesis* follows immediately from theorem 5.3.

To prove *base change*, the fourth Weil conjecture, one needs the theorems on smooth and proper base change in étale cohomology which we did not explain because we will not use them in this book.

Definition 5.6   Let k be a (not necessarily finite) field with separable closure $\bar{k}$ and absolute Galois group G_k. For all $n \in \mathbb{N}_+$, the group $\mu_{l^n} \subset \bar{k}$ of l^n-th roots of unity is cyclic of order l^n, and we get a projective system

$$\cdots \longrightarrow \mu_{l^4} \longrightarrow \mu_{l^3} \longrightarrow \mu_{l^2} \longrightarrow \mu_l$$

where the connecting morphisms are given by taking the l-th power. We denote the inverse limit of this system by $\mathbb{Z}_l(1)$.

It is easy to see that $\mathbb{Z}_l(1)$ is a free $\mathbb{Z}_l$-module of rank 1 which is *non-canonically* isomorphic to (the additive group of) $\mathbb{Z}_l$ — an isomorphism $\mathbb{Z}_l \xrightarrow{\sim} \mathbb{Z}_l(1)$ is given by sending $1 \in \mathbb{Z}_l$ to a compatible system $(\zeta_n) \in \mathbb{Z}_l(1)$ where each ζ_n is a *primitive* l^n-th root of unity.

The natural action of G_k on $\bar{k}^\times$ restricts to a G_k-action on μ_{l^n} for all $n \in \mathbb{N}_+$ which is compatible with the maps $\mu_{l^{n+1}} \to \mu_{l^n}$ and thus induces a $\mathbb{Z}_l$-linear G_k-action on $\mathbb{Z}_l(1)$. This action corresponds to a group homomorphism

$$\chi_l : G_k \longrightarrow \text{Aut}_{\mathbb{Z}_l}(\mathbb{Z}_l(1)) \xrightarrow{\sim} \mathbb{Z}_l^\times,$$

which is called the *l-adic cyclotomic character of G_k* (note that because the $\mathbb{Z}_l$-rank of $\mathbb{Z}_l(1)$ is one, the isomorphism $\mathrm{Aut}_{\mathbb{Z}_l}(\mathbb{Z}_l(1)) \cong \mathbb{Z}_l^\times$ is canonical, i.e. does not depend on the choice of a $\mathbb{Z}_l$-base in $\mathbb{Z}_l(1)$). Via

$$G_k \xrightarrow{\chi_l} \mathbb{Z}_l^\times \hookrightarrow \mathbb{Q}_l^\times = \mathrm{Aut}_{\mathbb{Q}_l}(\mathbb{Q}_l)$$

we get an induced $\mathbb{Q}_l$-linear G_k-action on $\mathbb{Q}_l$ which we also denote by χ_l. The resulting object $(\mathbb{Q}_l, \chi_l)$ of $\mathbf{Rep}_{\mathbb{Q}_l}^{G_k}$ is denoted by $\mathbb{Q}_l(1)$. — It is easy to see that χ_l is *continuous* with respect to the Krull-topology on G_k and the *l*-adic topology on $\mathbb{Q}_l$.

For arbitrary integers $d \in \mathbb{Z}$, we define objects $\mathbb{Q}_l(d)$ of $\mathbf{Rep}_{\mathbb{Q}_l}^{G_k}$ by

$$\mathbb{Q}_l(d) := \begin{cases} \mathbb{Q}_l(1)^{\otimes d} & \text{for } d \geq 0, \\ \mathrm{Hom}_{\mathbf{Rep}_{\mathbb{Q}_l}^{G_k}}(\mathbb{Q}_l(-d), \mathbb{Q}_l) & \text{for } d < 0. \end{cases}$$

Lemma 5.5 *Let $k = \mathbb{F}_q$ be a finite field, let $d \in \mathbb{Z}$ be an arbitrary integer, and let $f \in G_k$ be the arithmetic Frobenius $x \mapsto x^q$ of k. Then f acts on $\mathbb{Q}_l(d)$ by multiplication by q^d.*

Proof. By definition of $\mathbb{Q}_l(d)$, we only have to show that f acts on $\mathbb{Z}_l(1)$ by multiplication by q, but this is obvious, since $f(\zeta) = \zeta^q$ for a primitive l^n-th root of unity $\zeta \in \bar{k}$. $\square$

Remark 5.2 *Let l be a prime number, and let $\chi_l : G_\mathbb{Q} \to \mathbb{Q}_l^\times$ be the l-adic cyclotomic character defined in definition 5.6.*

As stated in the definition, χ_l is continuous with respect to the l-adic topology on $\mathbb{Q}_l$. But what happens if we endow $\mathbb{Q}_l$ with the discrete topology instead? — If χ_l was continuous as well then, the kernel of χ_l would have to be open in $G_\mathbb{Q}$, i.e. there would have to be a finite extension $K/\mathbb{Q}$ such that $\chi_l|G_K = 1$. But by construction of χ_l, it is obvious that the kernel of χ_l is the composite of all the cyclotomic fields $\mathbb{Q}(\zeta_{l^n})$ for $n \in \mathbb{N}_+$, and this composite has infinite degree over $\mathbb{Q}$ because $[\mathbb{Q}(\zeta_{l^n}) : \mathbb{Q}] = l^{n-1} \cdot (l - 1) \xrightarrow{n \to \infty} \infty$.

It follows that χ_l is not continuous with respect to the discrete topology on $\mathbb{Q}_l$.

Formula (5.9) or (5.11) allows us to compute the zeta function of the variety X, *provided* we know the characteristic polynomial of the geometric Frobenius on l-adic cohomology. In this book, we are mainly interested in *hypersurfaces* (namely those hypersurfaces that are given by forms of Fermat equations), and for hypersurfaces, the situation is easier, because for them only the *middle* l-adic cohomology is non-trivial, by which we mean the following:

Proposition 5.3   *Let $n \geq 2$ be a natural number, and let $P \in k[X_1, \ldots, X_n]$ be a homogenous polynomial for which $X := X(P)$, the $(n-2)$-dimensional projective hypersurface defined by P in $\mathbb{P}_k^{n-1}$, is smooth[1]. Put*

$$Q(P,T) := P_{n-2}^X(T) = \det\left(1 - F_X^* T \,\middle|\, H_{\acute{e}t}^{n-2}(\bar{X}, \mathbb{Q}_l)\right)$$

Then

$$\zeta(P,T) = Q(P,t)^{[(-1)^{n+1}]} \prod_{d \in \{0,\ldots,n-2\} \setminus \{\frac{n-2}{2}\}} \frac{1}{1 - q^d T} \qquad (5.12)$$

and

$$\forall i \in \mathbb{N}_+ : \nu_X^{(i)} = \left(\sum_{\substack{d \in \{0,\ldots,n-2\} \\ d \neq \frac{n-2}{2}}} q^{id} \right) + (-1)^n \cdot Tr\left((F^*)^i \,\middle|\, H_{\acute{e}t}^{n-2}(\bar{X}, \mathbb{Q}_l)\right). \qquad (5.13)$$

Proof.   Deligne proves in [DK73, XI] for a smooth, projective hypersurface X over k of dimension $(n-2)$ and for $d \in \{0, \ldots, 2(n-2)\} \setminus \{n-2\}$ that

$$H_{\acute{e}t}^d(X, \mathbb{Q}_l) = \begin{cases} 0 & \text{if } d \text{ is odd} \\ \mathbb{Q}_l\left(-\frac{d}{2}\right) & \text{if } d \text{ is even.} \end{cases}$$

By proposition 5.1 and lemma 5.5, the geometric Frobenius of X acts on

---

[1] This condition is in particular satisfied if P is a form of the Fermat equation of degree prime to p.

$\mathbb{Q}_l(-\frac{d}{2})$ by multiplication by $\frac{q}{2}$ and therefore

$$P_d^X(T) = \begin{cases} 1 & \text{if } d \text{ is odd} \\ 1 - q^{\frac{d}{2}}T & \text{if } d \text{ is even.} \end{cases}$$

So equation (5.12) follows from formula (5.11), and equation (5.13) follows from (5.9). □

Finally, we want to use l-adic cohomology to define a morphism of co-efficient extensions which will be vitally important for the computation of the zeta function of a twisted Fermat equation.

Example 5.1   Let $i \in \mathbb{N}_0$ be a natural number, and let K/k be a Galois extension of (not necessarily finite) fields with Galois group G.

(1) Let $F' : \mathbf{Var}_k^{\mathrm{Iso}} \to \mathbf{Var}_K^{\mathrm{Iso}}$ be the coefficient extension from example 3.2 respectively example 3.3(6), and let $F'' : \mathbf{Rep}_{\mathbb{Q}_l}^{G_k} \to \mathbf{Rep}_{\mathbb{Q}_l}^{G_K}$ be the coefficient extension from example 3.3(5).

We define functors $F_k' : \mathbf{Var}_k^{\mathrm{Iso}} \to \mathbf{Rep}_{\mathbb{Q}_l}^{G_k}$ and $F_K' : \mathbf{Var}_K^{\mathrm{Iso}} \to \mathbf{Rep}_{\mathbb{Q}_l}^{G_K}$ as follows: For a projective variety X/k, let $F_k'X := \mathrm{H}_{\text{ét}}^i(\bar{X}, \mathbb{Q}_l)$, and for an isomorphism $g : X \xrightarrow{\sim} X'$ in $\mathbf{Var}_k^{\mathrm{Iso}}$, we put $F_k'g := (g^*)^{-1}$. This obviously defines a *covariant* functor from $\mathbf{Var}_k^{\mathrm{Iso}}$ to $\mathbf{Rep}_{\mathbb{Q}_l}^{G_k}$. We construct the functor F_K' in exactly the same way by replacing k by K.

Note that even though l-adic cohomology is contravariant, we have taken the *inverse* to define F_k' and F_K' on morphisms to really get *co-variant* functors. This is the reason why we can not work with the full categories $\mathbf{Var}_k$ and $\mathbf{Var}_K$ but have to restrict ourselves to *isomorphisms*.

For each projective variety X/k we have a canonical isomorphism

$$X \times_k \bar{K} \xrightarrow[\sim]{f_X'} X_K \times_K \bar{K},$$

which induces a canonical isomorphism

$$\mathrm{H}_{\text{ét}}^i(\bar{X}, \mathbb{Q}_l) \xrightarrow{\sim} \mathrm{H}_{\text{ét}}^i(\overline{X_K}, \mathbb{Q}_l),$$

and the collection of these isomorphisms defines an isomorphism $f' : F_K'F' \to F''F_k'$ of functors. We claim that the triple (F_k', F_K', f') then defines a morphism from F' to F'' — let us check this quickly:

For that, let $Y, Z \in \mathrm{Ob}(\mathbf{Var}_k^{\mathrm{Iso}})$, $g : Y_K \xrightarrow{\sim} Z_K$ and $\bar{s} \in G$ be arbitrary, and let $s \in G_k$ be a lifting of $\bar{s}$. Then the following diagram obviously commutes:

$$
\begin{array}{ccc}
Y_K \times_K \bar{K} & \xrightarrow[\sim]{\left((1_Z \times \mathrm{Spec}\,\bar{s})^{-1} g (1_Y \times \mathrm{Spec}\,\bar{s})\right) \times 1_{\bar{K}}} & Z_K \times_K \bar{K} \\
f'_Y \Big\uparrow \wr & = & \wr \Big\uparrow f'_Z \\
Y \times_k \bar{K} & \xrightarrow[\sim]{(1_Z \times \mathrm{Spec}\,s)^{-1} (f'_Z g f'_Y{}^{-1})(1_Y \times \mathrm{Spec}\,s)} & Z \times_k \bar{K}
\end{array}
$$

So we get

$$
f' F'_K{}^{\bar{s}} g = \left[\mathrm{H}^i_{\text{ét}} \left((1_Z \times \mathrm{Spec}\,s^{-1})(f'_Z g f'_Y{}^{-1})(1_Y \times \mathrm{Spec}\,s) \right) \right]^{-1}
$$
$$
= \varphi_Z(s)(f' F'_k g)\varphi_Y(s)^{-1} = {}^{\bar{s}}(f' F'_k g),
$$

and this shows that the map

$$
\mathrm{Iso}_{\mathbf{Var}_K^{\mathrm{Iso}}}(F'Y, F'Z) \xrightarrow{\;f' F'_K\;} \mathrm{Iso}_{\mathbf{Rep}_{\mathbb{Q}_l}^{G_K}}(F'' F'_k Y, F'' F'_k Z)
$$

is indeed G-equivariant, in accordance with definition 3.3.

(2) Let F' and F'' be as above, let $n \geq 2$ and $r \geq 1$ be natural numbers, and let $\tilde{F} : \widetilde{\mathcal{F}_k^{n,r}} \to \widetilde{\mathcal{F}_K^{n,r}}$ be the coefficient extension from example 3.3(4). Remember that in example 3.4, we have defined a morphism $(\tilde{F}_k, \tilde{F}_K, \tilde{f})$ from $\tilde{F}$ to F'. Composing this morphism with the morphism from (1), we get a morphism $(H_k, H_K, h) : \tilde{F} \to F''$:

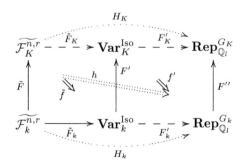

with $H_k := F'_k \tilde{F}_k$, $H_K := F'_K \tilde{F}_K$ and $h := (f' \tilde{F}_k) \circ (F'_k \tilde{f})$.

Chapter 6

Classification of forms

Throughout this chapter, let K/k be an arbitrary Galois extension of fields with Galois group $G := \mathrm{Gal}(K/k)$, and let $F : \mathcal{C}_k \to \mathcal{C}_K$ be a coefficient extension from k to K.

Let X and Y be objects of $\mathcal{C}_k$ that become isomorphic in $\mathcal{C}_K$, so that there is an isomorphism $FX \xrightarrow{\sim} FY$ in $\mathcal{C}_K$. In this case, we call Y (or rather the $\mathcal{C}_k$-isomorphism class $[Y]$ of Y) a $\mathcal{C}_K/\mathcal{C}_k$-*form of* X. In the special case that Y is already isomorphic to X in $\mathcal{C}_k$, we call Y a *trivial* $\mathcal{C}_K/\mathcal{C}_k$-form of X.

Definition 6.1   Let X be an object of $\mathcal{C}_k$. We define $E(\mathcal{C}_K/\mathcal{C}_k, X)$, the class[1] of $\mathcal{C}_K/\mathcal{C}_k$-*forms of* X, as follows:

$$E(\mathcal{C}_K/\mathcal{C}_k, X) := \Big\{ Y \in \mathrm{Ob}(\mathcal{C}_k) \,\Big|\, FY \cong_K FX \Big\} / \mathcal{C}_k\text{-isomorphisms}.$$

Let for example $k = \mathbb{R}$ and $K = \mathbb{C}$, and consider the coefficient extension $\mathcal{F}_{\mathbb{R}}^{1,2} \to \mathcal{F}_{\mathbb{C}}^{1,2}$ from example 3.3(3). The polynomials X^2 and $(-X^2)$ are non-isomorphic objects of $\mathcal{F}_{\mathbb{R}}^{1,2}$, because there is no real number whose square is (-1). But of course there is the complex number i with $i^2 = -1$, and therefore $i \in \mathbb{C}^* = \mathrm{GL}(1, \mathbb{C})$ defines an isomorphism from X^2 to $(-X^2)$ in $\mathcal{F}_{\mathbb{C}}^{1,2}$. Therefore (the $\mathbb{R}$-isomorphism class of) $(-X^2)$ is a non-trivial $\mathcal{F}_{\mathbb{C}}^{1,2}/\mathcal{F}_{\mathbb{R}}^{1,2}$-form of X^2.

---

[1]Note that we do not know yet whether $E(\mathcal{C}_K/\mathcal{C}_k, X)$ is a set or a proper class. We will prove that it is actually a set in theorem 6.1.

Now let us return to the general situation, and let X be an object of $\mathcal{C}_k$. By example 4.2(2), the automorphism group $\mathrm{Aut}_{\mathcal{C}_K}(FX)$ is a (discrete) G-group, so that the nonabelian cohomology $H^1(G, \mathrm{Aut}_{\mathcal{C}_K}(FX))$ is defined.

In the following theorem we will prove that we have a canonical injection from $E(\mathcal{C}_K/\mathcal{C}_k, X)$ into this cohomology — so we see that nonabelian cohomology can help us to classify $\mathcal{C}_K/\mathcal{C}_k$-forms.

Theorem 6.1 *Let X be an object of $\mathcal{C}_k$, and denote the discrete G-group $\mathrm{Aut}_{\mathcal{C}_K}(FX)$ by $A_F(X) := A(X)$. Let $[Y]$ be a $\mathcal{C}_K/\mathcal{C}_k$-form of X, so that there is an isomorphism $f : FY \xrightarrow{\sim} FX$ in $\mathcal{C}_K$. Then $\left[s \mapsto f^s(f^{-1})\right]$ is a (not necessarily continuous) 1-cocycle of G in $A(X)$, whose class in the nonabelian cohomology $H^1(G, A(X))$ only depends on $[Y]$ — we denote this cohomology class by $\vartheta[Y]$.*

The assignment $[Y] \mapsto \vartheta[Y]$ is injective; in particular this shows that $E(\mathcal{C}_K/\mathcal{C}_k, X)$ is a set (and not a proper class). We consider $E(\mathcal{C}_K/\mathcal{C}_k, X)$ as a pointed set by taking the trivial form $[X]$ as special element. In this way, we get a well defined, injective morphism of pointed sets

$$\vartheta : E(\mathcal{C}_K/\mathcal{C}_k, X) \hookrightarrow H^1(G, A(X)).$$

Proof.

- ϑ *well defined:* We have to show that $\vartheta[Y]$ is really a well defined cohomology class which does not depend on the choice of Y and f:

  - *1-cocycle:* For $s \in G$ put $a_s := f^s(f^{-1})$. Then we have:

$$a_{st} = f^{st}(f^{-1}) = f^s 1_{FY}{}^{st}(f^{-1}) \overset{(\mathrm{CE1})}{=} f\left[{}^s(f^{-1}){}^s f\right]{}^{st}(f^{-1})$$

$$\overset{(\mathrm{CE1})}{=} f^s(f^{-1})^s\left[f^t(f^{-1})\right] = a_s{}^s a_t.$$

  - *independent of the choice of f:* Suppose there are two isomorphisms
  $FY \underset{f'}{\overset{f}{\underset{\sim}{\rightrightarrows}}} FX$. We then get a commutative triangle

$$
\begin{array}{c}
FY \overset{f}{\underset{f'}{}}
\end{array}
$$

Putting $a_s := f^s(f^{-1})$ and $a'_s := f'^s(f'^{-1})$, we get:

$$a'_s = (b^{-1}f)^s(f^{-1}b) \stackrel{(\text{CE1})}{=} b^{-1}\left(f^s(f^{-1})\right){}^s b = b^{-1}a_s{}^s b \sim a_s,$$

i.e. (a_s) and (a'_s) are cohomologous cocycles which define the same element in $H^1(G, A(X))$.

- *independent of the choice of Y:* Let $[Y'] = [Y]$, and let $g : Y' \xrightarrow{\sim} Y$ be an arbitrary isomorphism in $\mathcal{C}_k$. Then $FY' \xrightarrow{Fg} FY \xrightarrow{f} FX$ is an isomorphism in $\mathcal{C}_K$, and it follows, that

$$(f[Fg])^s(f[Fg])^{-1} \stackrel{(\text{CE1})}{=} f[Fg] \underbrace{{}^s([Fg]^{-1})}_{\stackrel{(\text{CE2})}{=}[Fg]^{-1}} {}^s(f^{-1}) = f^s(f^{-1}).$$

- ϑ *injective:* Let $[Y]$ and $[Y']$ be two $\mathcal{C}_K/\mathcal{C}_k$-forms of X with $\vartheta[Y] = \vartheta[Y']$. Then we have to show that Y and Y' are already isomorphic in $\mathcal{C}_k$. Because FY and FY' are isomorphic in $\mathcal{C}_K$ (since both are isomorphic to FX), we have isomorphisms $FY' \xrightarrow{g} FY \xrightarrow{f} FX$ in $\mathcal{C}_K$. Then

$$\vartheta[Y'] = [(a'_s)_s] \text{ with } a'_s = fg^s(g^{-1}f^{-1}),$$
$$\vartheta[Y] = [(a_s)_s] \text{ with } a_s = f^s(f^{-1}).$$

By hypothesis, the 1-cocycles (a'_s) and (a_s) are cohomologous, so there is a $b \in A(X)$ with

$$\forall s \in G \quad : \quad a'_s = b^{-1}a_s{}^s b$$
$$\stackrel{(\text{CE1})}{\Leftrightarrow} fg^s(g^{-1})^s(f^{-1}) = b^{-1}f^s(f^{-1})^s b$$
$$\stackrel{(\text{CE1})}{\Leftrightarrow} \underbrace{{}^s(g^{-1})^s(f^{-1})^s(b^{-1})^s f}_{\stackrel{(\text{CE1})}{=}{}^s(g^{-1}f^{-1}b^{-1}f)}_{=h} = \underbrace{g^{-1}f^{-1}b^{-1}f}_{=:h}$$
$$\Leftrightarrow \quad {}^s h = h.$$

We thus have found an isomorphism $h : FY \xrightarrow{\sim} FY'$ which is fixed by the G-action on $\text{Iso}_{\mathcal{C}_K}(FY, FY')$. Then (CE2) tells us that $h = Fg$ for an isomorphism $g : Y \xrightarrow{\sim} Y'$ in $\mathcal{C}_k$, i.e. $[Y]$ and $[Y']$ indeed denote the same $\mathcal{C}_k$-isomorphism class.

- ϑ *morphism of pointed sets:* If we choose the identity on FX as isomorphism $f : FX \xrightarrow{\sim} FX$, then for all $s \in G$ we obviously have

$f^s(f^{-1}) = 1_{FX}$, which shows that under ϑ, the class $[X]$ is mapped to the trivial class in cohomology.

□

As a first application we want to show how to compute the automorphism group of a $\mathcal{C}_K/\mathcal{C}_k$-form Y of X in terms of $A(X)$ and $\vartheta[Y]$.

For this remember the notion of the "twist" of a G-group by a cocycle from lemma 4.11 and the fact that the isomorphism class of this twist only depends on the cohomology class of the twisting cocycle (by lemma 4.12).

Proposition 6.1   *Let X be an object of $\mathcal{C}_k$, and let $[Y]$ be a $\mathcal{C}_K/\mathcal{C}_k$-form of X. Then as a G-group, $A(Y)$ is isomorphic to the twisted group $A(X)_{\vartheta[Y]}$.*

If $FY \xrightarrow{\sim} FX$ is an isomorphism, then by definition $\vartheta[Y]$ is the class of the cocycle $a := (a_s) := (f^s(f^{-1}))$, and a G-isomorphism $A(Y) \to A(X)_a$ is explicitly given by $b \mapsto fbf^{-1}$.

If the map $\mathrm{Aut}_{\mathcal{C}_k}(Y) \xrightarrow{F} A(Y)$ is injective[2], then the automorphism group $\mathrm{Aut}_{\mathcal{C}_k}(Y)$ is isomorphic to the cohomology group $H^0(G, A(X)_{\vartheta[Y]})$.

Proof.   The map $b \mapsto fbf^{-1}$ is obviously a group isomorphism, so we only have to show that it is G-equivariant. For this, let $b \in A(Y)$ and $s \in G$ be arbitrary elements. We get

$$s'(fbf^{-1}) = a_s \cdot {}^s(fbf^{-1}) \cdot a_s^{-1} \overset{(CE1)}{=} f^s(f^{-1}) \cdot {}^sf^sb^s(f^{-1}) \cdot {}^sff^{-1}$$
$$= f \cdot {}^s(f^{-1}f) \cdot {}^sb \cdot {}^s(f^{-1}f) \cdot f^{-1} = f({}^sb)f^{-1}.$$

This shows that $A(Y)$ and $A(X)_{\vartheta[Y]}$ are really isomorphic G-groups. But then we consequently have $H^0(G, A(X)_{\vartheta[Y]}) \cong H^0(G, A(Y))$, and the latter group equals $\mathrm{Aut}_{\mathcal{C}_k}(Y)$ by (CE2) if $\mathrm{Aut}_{\mathcal{C}_k}(Y) \to A(Y)$ is injective. This concludes the proof of the proposition.

□

So far in this chapter, we have ignored the (Krull-)topology on G, because we have considered an arbitrary coefficient extension instead of a *continuous* one (see definition 3.4) and used arbitrary cohomology classes instead of continuous ones for our classification of forms.

---

[2]This condition is trivially satisfied for all coefficient extensions in example 3.3.

In the following lemma we show that if we restrict ourselves to *continuous* coefficient extension, then ϑ factorizes over *continuous* cohomology classes.

Lemma 6.1 *If the coefficient extension* $F : \mathcal{C}_k \to \mathcal{C}_K$ *is continuous, and if* X *is an object of* $\mathcal{C}_k$, *then* ϑ *factorizes over continuous cohomology classes:*

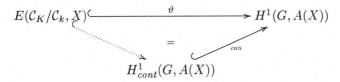

$$E(\mathcal{C}_K/\mathcal{C}_k, X) \xrightarrow{\quad\vartheta\quad} H^1(G, A(X))$$

$$H^1_{cont}(G, A(X))$$

Proof. Let $[Y]$ be an arbitrary $\mathcal{C}_K/\mathcal{C}_k$-form of X. Choose an isomorphism $f : FY \xrightarrow{\sim} FX$ in $\mathcal{C}_K$, and let $(a_s) = (f^s(f^{-1}))$ be the corresponding cocycle whose class is $\vartheta[Y]$. We have to show that (a_s) is continuous. By hypothesis, the set $\mathrm{Iso}_{\mathcal{C}_K}(FY, FX)$ is a discrete G-set, so according to lemma 4.1(1)$\Rightarrow$(2) we find an open subgroup U of G with $f \in \mathrm{Iso}_{\mathcal{C}_K}(FY, FX)^U$. For arbitrary $u \in U$ we have

$$1_{FX} = {}^u 1_{FX} = {}^u(ff^{-1}) = {}^u f \, {}^u(f^{-1}) = f \, {}^u(f^{-1}) = a_u.$$

So the cocycle (a_u) restricted to U is constant with value 1_{FX}, and for arbitrary $s \in G$ and $u \in U$ we get

$$a_{su} = a_s \, {}^s a_u = a_s \, {}^s 1_{FX} = a_s,$$

i.e. (a_s) is constant on the open subset sU of G for every $s \in G$. Therefore it is a locally constant and thus continuous function. $\qquad\square$

Let X be an object of $\mathcal{C}_k$. Thanks to theorem 6.1, we always have a canonical injection $E(\mathcal{C}_K/\mathcal{C}_K, X) \hookrightarrow H^1(G, A(X))$. However, for a complete classification of $\mathcal{C}_K/\mathcal{C}_k$-forms of X in terms of cohomology, we would like to know how to compute a form from its image under ϑ. Ideally, we would like ϑ to be bijective, and we would like to know ϑ^{-1} explicitly.

In general, ϑ will not be bijective, and even if it is, we will not have an explicit description of ϑ^{-1}.

Fortunately, in the cases of interest for us, we will not only be able to prove that ϑ is bijective, but will also be able to describe ϑ^{-1} explicitly.

As a first important example, in the next lemma we do this for the coefficient extension $F : \mathbf{Rep}_{\mathcal{C}}^{G_k} \to \mathbf{Rep}_{\mathcal{C}}^{G_K}$ from example 3.3(5).

Lemma 6.2   *Consider the special case where F is the coefficient extension $\mathbf{Rep}_{\mathcal{C}}^{G_k} \to \mathbf{Rep}_{\mathcal{C}}^{G_K}$ from example 3.3(5) for an arbitrary category $\mathcal{C}$. Let $X = (V, \varphi)$ be an object of $\mathbf{Rep}_{\mathcal{C}}^{G_k}$, and let $\xi = (a_{\bar{s}}) \in Z^1(G, A(X))$ be an arbitrary 1-cocycle of G in $A(X)$. Denote the image of an $s \in G_k$ under $G_k \twoheadrightarrow G$ by $\bar{s}$. Then*

$$\varphi^\xi : G_k \longrightarrow Aut_{\mathcal{C}}(V), \ s \mapsto \varphi^\xi(s) := a_{\bar{s}} \circ \varphi(s)$$

is a group homomorphism and thus defines a new object $X^\xi := (V, \varphi^\xi)$ of $\mathbf{Rep}_{\mathcal{C}}^{G_k}$. Furthermore we have that $[X^\xi]$ is a $\mathbf{Rep}_{\mathcal{C}}^{G_K}/\mathbf{Rep}_{\mathcal{C}}^{G_k}$-form of X with

$$\vartheta[X^\xi] \ = \ [\xi] \in H^1(G, A(X)).$$

In particular we see that ϑ is bijective in this case with ϑ^{-1} sending the class of a cocycle ξ to the class of X^ξ.

Proof.

- φ^ξ *group homomorphism:* For $s \in G_k$, both $a_{\bar{s}}$ and $\varphi(s)$ are automorphisms of V, so $\varphi^\xi(s)$ is also an automorphism of V, so that φ^ξ is a well defined *map*.

  Now let s and t be two arbitrary elements of G_k. We then get:

$$\varphi^\xi(st) = a_{\overline{st}} \circ \varphi(st) = a_{\bar{s}} \circ {}^{\bar{s}}a_{\bar{t}} \circ \varphi(st)$$
$$\overset{(3.6)}{=} a_{\bar{s}} \circ \left[\varphi(s) \circ a_{\bar{t}} \circ \varphi(s)^{-1} \right] \circ \left[\varphi(s) \circ \varphi(t) \right]$$
$$= \left[a_{\bar{s}} \circ \varphi(s) \right] \circ \left[a_{\bar{t}} \circ \varphi(t) \right] = \varphi^\xi(s) \circ \varphi^\xi(t).$$

- $\mathbf{Rep}_{\mathcal{C}}^{G_K}/\mathbf{Rep}_{\mathcal{C}}^{G_k}$-*form of X:* If s is an arbitrary element of $G_K \subseteq G_k$, then $\bar{s} = 1_G$, and consequently we get

$$\forall s \in G_K : \varphi^\xi(s) \overset{4.2}{=} 1_{A(X)} \cdot \varphi(s) = \varphi(s).$$

This shows that $(V, \varphi^\xi|_{G_K}) = (V, \varphi|_{G_K})$, so 1_V is a $\mathbf{Rep}_{\mathcal{C}}^{G_K}$-isomorphism from X^ξ to X and $[X^\xi]$ is a $\mathbf{Rep}_{\mathcal{C}}^{G_K}/\mathbf{Rep}_{\mathcal{C}}^{G_k}$-form of X.

- $\vartheta[X^\xi] = [\xi]$: As we have seen just now, as isomorphism $f : FX^\xi \xrightarrow{\sim} FX$ we can take the identity 1_V. It follows that $\vartheta[X^\xi] = [(b_{\bar{s}})]$ with

$$b_{\bar{s}} = 1_V \circ {}^{\bar{s}}(1_V^{-1}) \overset{(3.6)}{=} 1_V \circ \varphi^\xi(s) \circ 1_V \circ \varphi(s)^{-1} = a_{\bar{s}}\varphi(s)\varphi(s)^{-1} = a_{\bar{s}},$$

where s denotes a lift of $\bar{s}$ to G_k.

$\square$

Remark 6.1 *If G is finite, then ϑ is a bijection for the coefficient extensions from examples 3.2 and 3.3(2) as well (compare [Ser97, pp121]).*

Let (H_k, H_K, h) be a morphism from F to a coefficient extension $F' : \mathcal{C}_k' \to \mathcal{C}_{K'}$, and let X be an object of $\mathcal{C}_k$. Then we can ask the natural question: What is the relationship between $\mathcal{C}_K/\mathcal{C}_k$-forms of X and $\mathcal{C}_{K'}/\mathcal{C}_{k'}$-forms of $H_k X$, and how is this relationship reflected in the classifying cohomologies $H^1(G, A_F(X))$ and $H^1(G, A_{F'}(H_k X))$? — The answer is given in the next proposition:

Proposition 6.2 *Let (H_k, H_K, h) be a morphism from F to another coefficient extension $F' : \mathcal{C}_k' \to \mathcal{C}_K'$ from k to K, and let X be an object of $\mathcal{C}_k$. Then we have the following commuting diagram of pointed sets:*

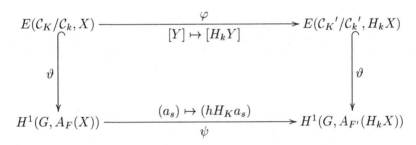

Proof. First we want to see that φ is well defined; for this we have to check that for $[Y] \in E(\mathcal{C}_K/\mathcal{C}_k, X)$, the $\mathcal{C}_k'$-isomorphism class $[H_k Y]$ lies indeed in $E(\mathcal{C}_{K'}/\mathcal{C}_k', K_k X)$, that it is independent of the choice of Y and that the trivial class $[X]$ is mapped to the trivial class $[H_k X]$.

So let Y be an object of $\mathcal{C}_k$, and let $f : FY \xrightarrow{\sim} FX$ be a $\mathcal{C}_K$-isomorphism. Then $H_K f$ is a $\mathcal{C}_K{}'$-isomorphism from $H_K(FY)$ to $H_K(FX)$, which implies that

$$F'(H_kY) \xrightarrow{h^{-1}} H_K(FY) \xrightarrow{H_K f} H_K(FX) \xrightarrow{h} F'(H_kX)$$

is a $\mathcal{C}_K{}'$-isomorphism from $F'(H_kY)$ to $F'(H_kX)$. Therefore $[H_kY]$ really is a $\mathcal{C}_K{}'/\mathcal{C}_k{}'$-form of H_kX.

Now let Y' be another object of $\mathcal{C}_k$ which is $\mathcal{C}_k$-isomorphic to Y, say via an $\mathcal{C}_k$-isomorphism $g : Y' \xrightarrow{\sim} Y$. Then $H_k g$ is a $\mathcal{C}_k{}'$-isomorphism from $H_k Y'$ to $H_k Y$, so φ is a well defined map *of sets*.

But obviously φ is also a morphism of *pointed sets*, because the trivial class is obviously mapped to the trivial class $[H_kX]$.

Next we want to check that ψ is well defined, i.e. we have to see that for a 1-cocycle (a_s) of G in $A(X)$, we get a 1-cocycle $(hH_K a_s)$ of G in $A(H_kX)$, and that for a cohomologous cocycle $(a_s') \sim (a_s)$ we have $(hH_K a_s') \sim (hH_K a_s)$.

Let $s, t \in G$ be arbitrary elements. It follows that

$$hH_K a_{st} = hH_K \left(a_s{}^s a_t\right) = (hH_K a_s)\left(hH_K{}^s a_t\right)$$
$$\overset{\text{Definition 3.3}}{=} (hH_K a_s)\,{}^s(hH_K a_t),$$

so $(hK_K a_s)$ satisfies the cocycle condition.

Now let (a_s') be a cocycle which is cohomologous to (a_s). By definition, there then is a $b \in A(X)$ such that $a_s' = b^{-1} a_s{}^s b$ for all $s \in G$. Then for all $s \in G$ we have

$$hH_K a_s' = hH_K \left(b^{-1} a_s{}^s b\right) = \underbrace{\left(hH_K b^{-1}\right)}_{= (hH_K b)^{-1}} (hH_K a_s) \underbrace{\left(hH_K{}^s b\right)}_{\overset{\text{Definition 3.3}}{=} {}^s(hH_K b)} ,$$

i.e. $(hH_K a_s')$ is cohomologous to $(hH_K a_s)$.

Finally, if (a_s) is the trivial cocycle in $Z^1(G, A(X))$, then obviously $(hH_K a_s)$ is the trivial cocycle of G in $A(H_kX)$, which shows that ψ is a well defined morphism of pointed sets.

It remains to show that the diagram commutes. For that, let $[Y]$ be an arbitrary $\mathcal{C}_K/\mathcal{C}_k$-form of X, and let $f : FY \xrightarrow{\sim} FX$ be an isomorphism in

$\mathcal{C}_K$. Then for all $s \in G$ we have

$$(\vartheta\varphi[Y])_s = (\vartheta[H_kY])_s = (hH_Kf)^s(hH_Kf^{-1})$$

$$\overset{\text{definition 3.3}}{=} hH_K \underbrace{(f^s(f^{-1}))}_{= (\vartheta[Y])_s} = (\psi\vartheta[Y])_s .$$

Thus the proof of the proposition is complete. $\qquad\square$

Corollary 6.1 *Consider the special case where F is a continuous coefficient extension, let (H_k, H_K, h) be a morphism from F to another continuous coefficient extension $F' : \mathcal{C}_k{}' \to \mathcal{C}_K{}'$ from k to K, and let X be an object of $\mathcal{C}_k$. Then we have the following commuting diagram of pointed sets:*

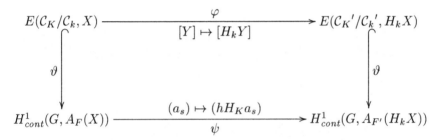

Proof. If $a : G \to A_F(X)$ is a continuous 1-cocycle, then the composition

$$G \overset{a}{\to} A_F(X) \overset{hH_K}{\longrightarrow} A_{F'}(X)$$

must also be continuous, because both $A_F(X)$ and $A_{F'}(X)$ carry the discrete topology. This shows that we get the following commutative diagram of pointed sets, from which the corollary immediately follows by lemma 6.1 and proposition 6.2.

$\qquad\square$

Let us have a closer look at the special case where F is the coefficient extension $\mathbf{Var}_k^{\mathrm{Iso}} \to \mathbf{Var}_K^{\mathrm{Iso}}$ from example 3.2 respectively example 3.3(6), where F' is the coefficient extension $\mathbf{Rep}_{\mathbb{Q}_l}^{G_k} \to \mathbf{Rep}_{\mathbb{Q}_l}^{G_K}$ from example 3.3(5) for a prime l different from the characteristic of k and an $i \in \mathbb{N}_0$, and where $F \to F'$ is the morphism given by l-adic cohomology defined in example 5.1(1).

For any variety X over k, we then have — according to proposition 6.2 — the following commuting diagram of pointed sets, where we write φ_Y for the l-adic representation $G_k \to \mathrm{H}^i_{\text{ét}}(Y \times_k \bar{K}, \mathbb{Q}_l)$ induced by a variety Y/k:

$$
\begin{array}{ccc}
E(\mathbf{Var}_K^{\mathrm{Iso}}/\mathbf{Var}_k^{\mathrm{Iso}}, X) & \xrightarrow[\ [Y] \mapsto [\varphi_Y]\ ]{\varphi} & E(\mathbf{Rep}_{\mathbb{Q}_l}^{G_K}/\mathbf{Rep}_{\mathbb{Q}_l}^{G_k}, \varphi_X) \\
\vartheta \Big\downarrow & = & \wr \Big\downarrow \vartheta \quad \xi \mapsto \varphi_X^\xi \\
H^1(G, A(X)) & \xrightarrow[\psi]{(a_s) \mapsto \left(\mathrm{H}^i_{\text{ét}}(a_s)^{-1}\right)} & H^1(G, A(\varphi_X))
\end{array}
$$

$$(6.1)$$

Note that from lemma 6.2, we know that the right ϑ is an isomorphism whose inverse we know explicitly.

What does that mean? — Suppose we have a $\mathbf{Var}_K^{\mathrm{Iso}}/\mathbf{Var}_k^{\mathrm{Iso}}$-form Y of X of which we know a cocycle (a_s) representing $\vartheta[Y]$, and suppose we know the map $A(X) \to A(\varphi_X)$ explicitly. Then diagram (6.1) tells us how to compute φ_Y, the l-adic representation of G_k induced by Y, from this data.

Example 6.1 Consider the special case where k is a number field, $K := \bar{\mathbb{Q}}$ is the field of algebraic numbers and X is the curve given by $X_1^3 + X_2^3 + X_3^3 = 0$. As we have seen in example 3.3(4), X is isomorphic to the curve with equation $12X_1^3 + X_2^2 X_3 + 3X_3^3 = 0$. Setting $X_1 := x$, $X_2 := y$ and $X_3 := 1$, we see that this is the elliptic curve with Weierstraß equation $y^2 = -12x^3 - 3$ over k. The j-invariant of this curve is 0, and it is well-known that two elliptic curves over $\bar{\mathbb{Q}}$ are isomorphic if and only if they have the same j-invariant (compare [Kna92, p66]). This means that an elliptic curve Y over k is a $\mathbf{Var}_K/\mathbf{Var}_k$-form of X if and only if it has j-invariant 0. We thus get

$$\left\{Y/k \text{ elliptic curve with } j\text{-invariant } 0\right\}/k\text{-isomorphisms}$$

$$= \left\{ [Y] \in E(\mathbf{Var}_K/\mathbf{Var}_k, X) \mid Y \text{ has a } k\text{-rational point} \right\}.$$

In principal, if we know the G_k-representation $\mathrm{H}^i_{\text{ét}}(\bar{X}, \mathbb{Q}_l)$, proposition 6.2 therefore enables us to compute the G_k-representation $\mathrm{H}^i_{\text{ét}}(\bar{Y}, \mathbb{Q}_l)$ for every elliptic curve Y/k with j-invariant 0. In particular, it enables us to compute the L-series of Y.

The same argument as above applies if we compose the morphism $F \to F'$ with a morphism $\tilde{F} \to F$; in particular we can take the morphism

$$\left(\widetilde{\mathcal{F}^{n,r}_k} \to \widetilde{\mathcal{F}^{n,r}_K} \right) \longrightarrow \left(\mathbf{Rep}^{G_k}_{\mathbb{Q}_l} \to \mathbf{Rep}^{G_K}_{\mathbb{Q}_l} \right)$$

from example 5.1(2) for natural numbers $n \geq 2$ and $r \geq 1$. We will make this precise in the following corollary:

Corollary 6.2   *Let $n \geq 2$, $r \geq 1$ and $i \geq 0$ be natural numbers, let l be a prime different from the characteristic of k, and let P be an object of $\widetilde{\mathcal{F}^{n,r}_k}$ — remember that this means that P is a homogenous polynomial of degree r in $k[X_1, \ldots, X_n]$.*

Denote the hypersurface in $\mathbb{P}^{n-1}_k$ corresponding to P by X, and let φ_P be the action of G_k on the l-adic cohomology V of X:

$$\varphi_P : G_k \longrightarrow \mathrm{Aut}_{\mathbb{Q}_l}(\underbrace{H^i_{\text{ét}}(X \times_k \bar{K}, \mathbb{Q}_l)}_{=:V}).$$

Take a $\widetilde{\mathcal{F}^{n,r}_K}/\widetilde{\mathcal{F}^{n,r}_k}$-Form Q of P with $\vartheta[Q] =: [(A_s)] \in H^1(G, A(P))$ and corresponding projective hypersurface Y. Because the varieties Y_K and X_K are isomorphic over K, the cohomology $H^i_{\text{ét}}(Y \otimes_k \bar{K}, \mathbb{Q}_l)$ is — as a $\mathbb{Q}_l$-vector space — isomorphic to V. Then φ_Q, the G_k-action on V induced by Q respectively Y, is given by

$$\varphi_Q : G_k \longrightarrow \mathrm{Aut}_{\mathbb{Q}_l}(V), \quad s \mapsto H^i_{\text{ét}}\left(A_{\bar{s}}^{-1}\right) \circ \varphi_P(s) \tag{6.2}$$

In the particular case that k is finite with arithmetic Frobenius $f \in G_k$, we get:

$$F^*_Y = H^i_{\text{ét}}\left(\bar{f}^{-1} A_{\bar{f}}\right) \circ F^*_X \tag{6.3}$$

where $F^*_P := \varphi_P(f)^{-1}$ and $F^*_Q := \varphi_Q(f)^{-1}$ denote the geometric Frobenii. In this case, the knowledge of F^*_Q uniquely determines φ_Q.

Proof.   Equation (6.2) is an easy consequence of lemma 6.2 and proposition 6.2 (respectively diagram (6.1)). To deduce equation (6.3) from equation (6.2), we only have to show that $A_{\bar{f}^{-1}}^{-1} = {}^{\bar{f}^{-1}}A_{\bar{f}}$:

$$1_{\mathrm{PGL}\,(n,k)} \overset{\text{lemma 4.2}}{=} A_{1_G} = A_{\bar{f}^{-1}\bar{f}} = A_{\bar{f}^{-1}} \cdot {}^{\bar{f}^{-1}}A_{\bar{f}}$$

$$\implies \quad A_{\bar{f}^{-1}}^{-1} = {}^{\bar{f}^{-1}}A_{\bar{f}}.$$

φ_Q is continuous with respect to the Krull-topology on G_k and the l-adic topology on V, and f (and hence f^{-1}) generate G_k topologically, so the knowledge of F_Q^* uniquely determines φ_Q.   □

Chapter 7

Forms of the Fermat equation I

In this chapter, let r and n be positive natural numbers, let k be a field whose characteristic is greater than r, let $K := \bar{k}$ be a separable closure of k, and let $G := \mathrm{Gal}(K/k)$ be the absolute Galois group of k.

Denote the Fermat equation $X_1^r + \ldots + X_n^r \in k[X_i]$ by P_n^r. It is an object of $\mathcal{F}_k^{n,r}$ and of $\widetilde{\mathcal{F}_k^{n,r}}$, but in this chapter we only want to consider it as an object of $\mathcal{F}_k^{n,r}$, and we want to determine its $\mathcal{F}_K^{n,r}/\mathcal{F}_k^{n,r}$-forms.

As we know from the last chapter, we first have to compute the group of automorphisms $A(P_n^r)$ in order to be able to apply our cohomological tools to determine those forms.

We start with a lemma about wreath products (compare definition 4.8):

Lemma 7.1 *Let m be a positive natural number, let R be a commutative ring with unit, and let $A \subseteq R^\times$ be a subgroup of the group of units of R. Then we get an injective group homomorphism from the wreath product $A \wr S_m$ into the group of invertible $m \times m$-matrices over R by*

$$A \wr S_m \;\hookrightarrow\; GL(m, R)$$
$$(a_i)_i \cdot \sigma \;\mapsto\; \left(a_i \cdot \delta_{i, \sigma(j)} \right)_{ij}.$$

Under this map, elements of S_r are mapped to permutation matrices, and elements of A^r are mapped to diagonal matrices.

Proof. Denote the map by f.

- *multiplicative:* Let $(a_i)_i \cdot \sigma$ and $(b_i)_i \cdot \tau$ be arbitrary elements of $A \wr S_m$.

105

Then we get:

$$f\big[(a_i)_i \cdot \sigma\big] \cdot f\big[(b_i)_i \cdot \tau\big]$$
$$= \big(a_i \cdot \delta_{i,\sigma(k)}\big)_{i,k} \cdot \big(b_k \cdot \delta_{k,\tau(j)}\big)_{k,j} \;=\; \big(\textstyle\sum_{k=1}^{m} a_i \cdot \delta_{i,\sigma(k)} \cdot b_k \cdot \delta_{k,\tau(j)}\big)_{i,j}$$
$$= \big(a_i \cdot \delta_{i,\sigma\tau(j)} \cdot b_{\tau(j)}\big)_{i,j} \;=\; \big(a_i b_{\sigma^{-1}(i)} \cdot \delta_{i,\sigma\tau(j)}\big)_{i,j}$$
$$= f\big((a_i b_{\sigma^{-1}(i)})_i \cdot \sigma\tau\big) \;=\; f\big([(a_i)_i \cdot \sigma] \cdot [(b_i)_i \cdot \tau]\big).$$

- *well defined:* We have $f(1_{A \wr S_m}) = f[(1, \dots, 1) \cdot 1_{S_m}] = 1_{\mathrm{GL}\,(m,R)}$, so for arbitrary $x \in A \wr S_m$ the fact that f is multiplicative implies $f(x) \cdot f(x^{-1}) = f(x^{-1}) \cdot f(x) = f(1_{A \wr S_m}) = 1_{\mathrm{GL}\,(m,R)}$, so $f(x)$ is indeed invertible.
- *injective:* If an $(a_i)_i \cdot \sigma$ is an arbitrary element of the kernel of f, we have for all $i \in \{1, \dots, m\}$ that $\delta_{ij} = \delta_{i,\sigma(j)}$, i.e. $i = \sigma(i)$ and therefore $\sigma = 1_{S_m}$. This shows that $a_i = 1$ for all i, so that $(a_i)_i \cdot \sigma = (1, \dots, 1) \cdot 1_{S_m} = 1_{A \wr S_m}$. $\qquad\square$

The significance of the wreath product for us is the following proposition which states that the automorphism group $A(P_n^r)$ is a particular wreath product:

Proposition 7.1 *If $r \geq 3$, then*

$$A(P_n^r) = Aut_{\mathcal{F}_K^{n,r}}(P_n^r) = \mu_r \wr S_n \subseteq GL\,(n, K),$$

where $\mu_r \subseteq K^\times$ is the group of r-th roots of unity in K. Here we consider the wreath product as a subgroup of $GL\,(n, K)$ via the embedding defined in lemma 7.1, i.e. we get an induced isomorphism as in the following diagram:

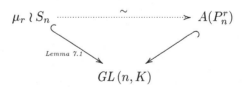

Proof. It is easy to see that the elements of $\mu_r \wr S_n$ define automorphisms of P_n^r over K (also for $r < 3$): The polynomial P_n^r equals $x_{\sigma^{-1}(1)}^r + \dots + x_{\sigma^{-1}(n)}^r$ for any $\sigma \in S_n$, and it also equals $[\zeta x_1]^r + \dots + [\zeta x_n]^r$ for any r-th root of unity ζ. The proof that these are *all* automorphisms for $r \geq 3$ can be found in [Shi88]. $\qquad\square$

Both $\mu_r \wr S_n$ and $A(P_n^r)$ are discrete G-groups: The continuous G-action on $K^\times$ induces a continuous G-action on $\mu_r \wr S_n$ according to lemma 4.8, given by equation (4.6), and $A(P_n^r)$ is a discrete G-group via equation (3.5), as proven in example 3.3(3) and example 3.5(3). The next lemma states that for $r \geq 3$, the isomorphism of proposition 7.1 is compatible with these structures:

Lemma 7.2 *For $r \geq 3$, the isomorphism of proposition 7.1 is an isomorphism of discrete G-groups with respect to the structures defined by equations (4.6) and (3.5).*

Proof. Assume $r \geq 3$, and denote the isomorphism $\mu_r \wr S_n \xrightarrow{\sim} A(P_n^r)$ from proposition 7.1 by f. Let $(a_i)_i \cdot \sigma$ be an arbitrary element of $\mu_r \wr S_n$, and let s be an arbitrary element of G. Then

$$f\left({}^s[(a_i)_i \cdot \sigma]\right) \overset{(4.6)}{=} f\left[({}^s a_i)_i \cdot \sigma\right] = \left({}^s a_i \cdot \delta_{i,\sigma(j)}\right)_{ij} = \left({}^s a_i \cdot {}^s \delta_{i,\sigma(j)}\right)_{ij}$$

$$= \left({}^s[a_i \cdot \delta_{i,\sigma(j)}]\right)_{ij} \overset{(3.5)}{=} {}^s\left[(a_i \cdot \delta_{i,\sigma(j)})_{ij}\right] = {}^s\left(f[(a_i)_i \cdot \sigma]\right).$$

Note that $\delta_{i,\sigma(j)}$ is fix under G because of $\{0,1\} \subseteq k$. $\qquad\square$

Now that we have determined the group of automorphisms $A(P_n^r)$, we can start applying our methods from chapter 4 to the computation of the cohomology $H^1_{\mathrm{cont}}(G, A(P_n^m))$. This will then enable us to calculate the $\mathcal{F}_K^{n,r}/\mathcal{F}_k^{n,r}$-Forms of P_n^r using the results from the last chapter.

Remember that for a cocycle $c \in Z^1_{\mathrm{cont}}(G, S_n)$ and a discrete G-group A, we have the exact sequence

$$1 \to A^n \to A \wr S_n \to S_n \to 1$$

from (4.7) and thus can consider the twisted group A_c^n — we explicitly described the G-action on A_c^n in equation (4.14).

Setting $A := \mu_r$ respectively $A := K^\times$ we get the discrete G-groups $(\mu_r)_c^n$ and $(K^\times)_c^n$ for which we have the following lemma:

Lemma 7.3 *For a cocycle $c \in Z^1_{cont}(G, S_n)$, the following short exact sequence of discrete G-groups is exact:*

$$1 \longrightarrow (\mu_r)_c^n \longrightarrow (K^\times)_c^n \overset{r}{\longrightarrow} (K^\times)_c^n \longrightarrow 1. \qquad (7.1)$$

Proof.   Because K is separably closed, the sequence

$$1 \to \mu_r \to K^\times \overset{r}{\to} K^\times \to 1$$

is an exact sequence of abelian groups, and taking the n-fold direct sum of this sequence, we see that sequence (7.1) is also an exact sequence of abelian groups.

In addition to that, equation (4.14) immediately shows that all occurring morphisms are morphisms of discrete G-groups, so the lemma is proven. $\square$

Hilbert's famous "Theorem 90" states that the Galois cohomology group $H^1_{\mathrm{cont}}(G, K^\times)$ is trivial. The following lemma is a generalization of this fact to the twisted G-group $(K^\times)^n_c$:

Lemma 7.4   *(Hilbert 90)*
Let $c \in Z^1_{cont}(G, S_n)$ be an arbitrary cocycle. Then

$$H^1_{cont}(G, (K^\times)^n_c) = 0.$$

This statement remains true if we replace K by an arbitrary Galois extension of k, and we will prove it for this more general situation.

Proof.   This proof closely follows [Rup03, 4.9, p29] and is a generalization of Serre's proof for the classical Theorem 90 in [Ser79, X.1, p150].

We are in the situation of lemma 4.17 whose notation we are going to use. In particular, c is given by an fcs-algebra L of degree n over k, and we put $M := \mathrm{Hom}_k(L, K)$ and identify S_n with $\mathrm{Aut}(M)$ and $(K^\times)^n$ with $(K^\times)^M$. Because of lemma 4.18 and lemma 4.5, we can assume without loss of generality that L is a *field*.

Let $\bar{L}$ be the normal hull of L in K, and let N be an open subgroup of G with $N \leq \mathrm{Gal}(K/\bar{L})$, corresponding to a finite Galois extension $K' = K^N$ of k containing $\bar{L}$. Which elements of $(K^\times)^n_c$ are fix under N? — If $(x_\varphi)_\varphi$ is an element of $(K^\times)^n_c$, and if $s \in N$, then

$$ {}^{s'}(x_\varphi)_\varphi \overset{(4.14)}{=} ({}^s x_{s^{-1} \circ \varphi})_\varphi \overset{s \in \mathrm{Gal}(K/\bar{L})}{=} ({}^s x_\varphi)_\varphi. $$

This shows that $(x_\varphi)_\varphi$ is fix under N if and only if each x_φ is fix under N, i.e. if and only if each x_φ is an element of K'. We thus have seen that

$[(K^\times)^n_c]^N = (K'^\times)^n_c$ and can now conclude from corollary 4.8(2):

$$H^1_{\mathrm{cont}}(G,(K^\times)^n_c) = \varinjlim_{N \in \mathcal{N}_G} H^1_{\mathrm{cont}}(G/N,[(K^\times)^n_c]^N)$$

$$= \varinjlim_{\substack{N \in \mathcal{N}_G \\ N \leq \mathrm{Gal}(K/\bar{L})}} H^1_{\mathrm{cont}}(G/N,([K^N]^\times)^n_c),$$

which means that (by replacing G with G/N and K with K^N) without loss of generality we can assume that G is *finite*.

We will need the following well known result due to Artin on the linear independence of characters (for a proof, see [Lan93, VI,§4.1, p283]):

> Let H be a monoid and F a field. Let $\chi_1,\ldots,\ldots,\chi_m$ be distinct homomorphisms $H \to F^\times$. Then they are linearly independent over F when considered as functions from H to F.

Let $(a_s)_s = ([a_s^{(\varphi)}]_\varphi)_s$ be a 1-cocycle from G in $(K^\times)^n_c$, and let φ_0 be an element of M. We will apply Artin's theorem to the situation where $H := K^\times$, $F := K$, $m := [K:k]$ and where for $s \in G$ the homomorphism χ_s is just s restricted to $K^\times$. Then because the $a_s^{(\varphi_0)}$ are not zero, the theorem tells us that the function $\sum_{s \in G} a_s^{(\varphi_0)} \chi_s$ from $K^\times$ to K is not zero, so we find a $c \in K^\times$ satisfying

$$\sum_{s \in G} a_s^{(\varphi_0)} \cdot {}^s c \in K^\times. \tag{7.2}$$

Put

$$b := (b_\varphi)_\varphi := \sum_{s \in G} a_s \cdot \underbrace{{}^s(c,\ldots,c)}_{n\text{-times}} \in K^n.$$

We claim that b is actually an element of $(K^\times)^n$. To see this, let φ be an arbitrary element of M. Because G acts transitively on M, there is a $t_\varphi \in G$ with $\varphi = t_\varphi \circ \varphi_0$. We get:

$$b_\varphi = \sum_{s \in G} a_s^{(\varphi)} \cdot {}^s c = \sum_{s \in G} a_{t_\varphi \cdot s}^{(\varphi)} \cdot {}^{t_\varphi \cdot s} c = \sum_{s \in G} a_{t_\varphi}^{(\varphi)} \cdot {}^{t_\varphi} a_t^{(t_\varphi^{-1} \circ \varphi)} \cdot {}^{t_\varphi \cdot s} c$$

$$= \sum_{s \in G} a_{t_\varphi}^{(\varphi)} \cdot {}^{t_\varphi} a_t^{(\varphi_0)} \cdot {}^{t_\varphi \cdot s} c = a_{t_\varphi}^{(\varphi)} \cdot {}^{t_\varphi} \underbrace{\left(\sum_{s \in G} a_t^{(\varphi_0)} \cdot {}^s c \right)}_{\substack{(7.2) \\ \neq 0}} \neq 0,$$

110   *Forms of Fermat Equations and their Zeta Functions*

so b is a well defined element of $(K^\times)^n_c$. We want to show that $(a_s)_s = (b \cdot {}^s(b^{-1}))_s$ because this implies that $(a_s)_s$ is cohomologous to the trivial cocycle, now that we know that $b \in (K^\times)^n_c$. Equivalently, we can show that $a_s \cdot {}^s b = b$ for all $s \in G$, but this is purely formal:

$$a_s \cdot {}^s b = \sum_{t \in G} \underbrace{a_s \cdot {}^s a_t}_{=a_{st}} \cdot {}^{st}(c,\dots,c) = \sum_{t \in G} a_t \cdot {}^t(c,\dots,c) = b.$$

$\square$

Lemma 7.5   *Let $c \in Z^1_{cont}(G, S_n)$ be a cocycle. Then with the notation of lemma 4.17, we have the following isomorphism of abelian groups (and so in particular of pointed sets):*

$$L^\times \xrightarrow{\sim} H^0(G, (K^\times)^n_c)$$
$$x \mapsto (\varphi x)_{\varphi \in M}.$$

Proof.   Denote the map in question by f, and for $i \in \{1,\dots,m\}$ denote the canonical map $L \twoheadrightarrow L_i \subseteq K$ by ι_i — the maps $\iota_1,\dots,\iota_m$ are therefore elements of M.

We claim that the inverse g of f is given by:

$$H^0(G, (K^\times)^n_c) \xrightarrow{g} \prod_{i=1}^m L_i^\times = L^\times$$
$$(x_\varphi)_{\varphi \in M} \mapsto (x_{\iota_i})_i.$$

First we want to check that g is well defined, i.e. that for arbitrary $(x_\varphi)_\varphi \in H^0(G, (K^\times)^n_c))$ and arbitrary $i \in \{1,\dots,m\}$ we have $x_{\iota_i} \in L_i$. For this we have to show that x_{ι_i} is fixed by the action of $G_{L_i} := \mathrm{Gal}(K/L_i)$.

So let s be an arbitrary element of G_{L_i}. Since G_{L_i} is a subgroup of G, by hypothesis (x_φ) is fixed by s, and we get

$$(x_\varphi)_\varphi = {}^s(x_\varphi)_\varphi \overset{(4.14)}{=} ({}^s x_{s^{-1}\varphi})_\varphi \implies x_{\iota_i} = {}^s x_{s^{-1}\iota_i} \overset{s|_{L_i}=1_{L_i}}{=} {}^s x_{\iota_i}.$$

Next we want to see that f itself is well defined. For this let $x \in L^\times$ and $s \in G$ be arbitrary. Then it follows:

$${}^s(\varphi x)_\varphi \overset{(4.14)}{=} ({}^s([s^{-1}\varphi]x))_\varphi = (ss^{-1}\varphi x)_\varphi = (\varphi x)_\varphi,$$

so $f(x) = (\varphi x)_\varphi$ is indeed an element of $H^0(G, (K^\times)^n_c)$.

Since f and g are obviously group homomorphisms, all that remains to be seen is that f and g are inverse to each other. First let $x = (x_i)_i$ be an arbitrary element of $L^\times$. Then

$$(gf)(x) = g\big[(\varphi x)_\varphi\big] = (\iota_i x)_i = (x_1, \ldots, x_m) = x.$$

Now let $(x_\varphi)_\varphi$ be an element of $H^0(G, (K^\times)^n_c)$. We have to show that for all $\psi \in M$

$$x_\psi = \Big((fg)\big[(x_\varphi)_\varphi\big]\Big)_\psi = \Big(f(x_{\iota_1}, \ldots, x_{\iota_m})\Big)_\psi = \psi(x_{\iota_1}, \ldots, x_{\iota_m}).$$

So let $\psi : L \to K$ be an arbitrary element of M. Then there is $j \in \{1, \ldots, m\}$ such that ψ factorizes over L_j, and consequently we have a $s \in G$ with $\psi = s\iota_j$. With this we get:

$$(x_\varphi)_\varphi \overset{(4.14)}{=} {}^s(x_\varphi)_\varphi = ({}^s x_{s^{-1}\varphi})_\varphi$$
$$\implies \quad x_\psi = {}^s x_{s^{-1}\psi} = {}^s x_{s^{-1}s\iota_j} = {}^s x_{\iota_j}$$
$$= {}^s(\iota_j(x_{\iota_1}, \ldots, x_{\iota_m})) = \psi(x_{\iota_1}, \ldots, x_{\iota_m}).$$

$$\square$$

Corollary 7.1   Let $c \in Z^1_{cont}(G, S_n)$ be an arbitrary cocycle. Then, using the notation from lemma 4.17, we have the following isomorphism of abelian groups:

$$L^\times / L^{\times^r} \overset{\gamma}{\to} H^1_{cont}(G, (\mu^n_r)_c)$$
$$\bar{x} \qquad \mapsto \qquad \left(\left(\frac{s\sqrt[r]{s^{-1}\varphi x}}{\sqrt[r]{\varphi x}}\right)_{\varphi \in M}\right)_s.$$

(For $y \in K^\times$, by $\sqrt[r]{y}$ we denote an arbitrary preimage of y under the surjection $K^\times \xrightarrow{z \mapsto z^r} K^\times$.)

Proof.   Taking the long exact cohomology sequence associated to the short exact sequence (7.1) and using lemma 7.5 and lemma 7.4, we get the following commutative diagram of abelian groups with exact columns:

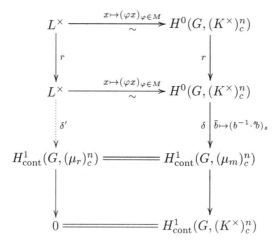

Thus the morphism δ' induces an isomorphism $\gamma \; : \; L^\times/L^{\times^r} \xrightarrow{\sim} H^1_{\text{cont}}(G, (\mu_r)^n_c)$, and all we have to do is prove that γ is given by the formula stated in the corollary. For this, let $\bar{x}$ be an arbitrary element of $L^\times/L^{\times^r}$, represented by an $x \in L^\times$. Then we get:

$$\gamma(\bar{x}) = \delta'(x) = \delta\big[(\varphi x)_\varphi\big] = \left(\left(\frac{1}{\sqrt[r]{\varphi x}}\right)_\varphi \cdot {}^s\big(\sqrt[r]{\varphi x}\big)_\varphi\right)_s$$

$$= \left(\left(\frac{s\sqrt[r]{s^{-1}\varphi x}}{\sqrt[r]{\varphi x}}\right)_\varphi\right)_s,$$

and the claim follows.   □

Corollary 7.2   *Let $b \in Z^1_{cont}(G, \mu_r \wr S_n)$ be an arbitrary cocycle. Then, using the notation from lemma 4.17 and lemma 4.18, we have the following isomorphism of abelian groups:*

$$\eta : \prod_{i=1}^{m} (L_i \cap \mu_r) \xrightarrow{\sim} H^0(G, (\mu_r^n)_c)$$
$$x \qquad\qquad \mapsto \quad (\varphi x)_{\varphi \in M}.$$

Proof.   Let c denote the image of b in $H^1_{\text{cont}}(G, S_n)$. According to proposition 4.2(4), the twisted G-actions on μ_r^n given by b and c coincide. Accordingly we may replace b by c. Because of lemma 7.3 and lemma 7.5 we then get the following commutative diagram of abelian groups with exact rows:

$$
\begin{array}{ccccccc}
1 & \longrightarrow & H^0(G, (\mu_r^n)_c) & \longrightarrow & H^0(G, (K^\times)_c^n) & \overset{r}{\longrightarrow} & H^0(G, (K^\times)_c^n) \\
& & \uparrow \wr \mid \eta & & \wr \Big\uparrow {\scriptstyle x \mapsto (\varphi x)_\varphi} & & \wr \Big\uparrow {\scriptstyle x \mapsto (\varphi x)_\varphi} \\
1 & \longrightarrow & \{x \in L^\times \mid x^r = 1\} & \longrightarrow & L^\times & \overset{r}{\longrightarrow} & L^\times
\end{array}
$$

So we get an isomorphism η by sending x to $(\varphi x)_\varphi$, and the corollary follows since we obviously have $\{x \in L^\times \mid x^r = 1\} = \prod_{i=1}^r (L_i \cap \mu_r)$. □

Lemma 7.6 *Let $c \in Z^1_{cont}(G, S_n)$ be an arbitrary cocycle. Then, using the notation from lemma 4.17, we have the following isomorphism of groups:*

$$
\begin{array}{rcl}
Aut_k(L)^{opp} & \overset{\alpha}{\to} & H^0(G, (S_n)_c) \\
a & \mapsto & [M \to M, \; \varphi \mapsto \varphi \circ a].
\end{array}
$$

Proof. Let $\iota_1, \ldots, \iota_m \in M$ be as in the proof of lemma 7.5. Using these, we claim that the inverse β to α is given as follows:

$$
\begin{array}{rcl}
H^0(G, (S_n)_c) & \overset{\beta}{\to} & Aut_k(L)^{opp} \\
\tau & \mapsto & \left(L \xrightarrow{(\tau(\iota_1), \ldots, \tau(\iota_m))} \prod_{i=1}^m L_i = L\right).
\end{array}
$$

Step by step we now want to prove that α and β are well defined, mutually inverse group isomorphisms:

- $\alpha(a)$ *bijective:* Let a be an arbitrary element of $Aut_k(L)^{opp}$. Then $\alpha(a)$ is obviously injective, since $\varphi a = \psi a$ implies $\varphi = \psi$ because a is bijective. But M is finite, so $\alpha(a)$ must bijective since it is injective.
- α *homomorphism:* Let $a, b \in Aut_k(L)^{opp}$ and $\varphi \in M$ be arbitrary elements. Then

$$
\begin{aligned}
\alpha(ab)(\varphi) = \alpha(b \circ a)(\varphi) = \varphi ba &= (\varphi b)a \\
&= \alpha(a)[\alpha(b)(\varphi)] = [\alpha(a) \circ \alpha(b)](\varphi).
\end{aligned}
$$

Thus we have $\alpha(ab) = \alpha(a) \circ \alpha(b)$.

- $\alpha(a) \in H^0(G, (S_n)_c)$: Let $a \in \mathrm{Aut}_k(L)^{\mathrm{opp}}$, $s \in G$ and $\varphi \in M$ be arbitrary elements. We compute

$$\left({}^s[\alpha(a)]\right)(\varphi) \overset{(4.14)}{=} s \circ \underbrace{\alpha(a)(s^{-1}\varphi)}_{=s^{-1}\varphi a} = ss^{-1}\varphi a = \varphi a = [\alpha(a)](\varphi)$$

and see, that we indeed have ${}^s[\alpha(a)] = \alpha(a)$. To sum up, we now know that α is a well defined group homomorphism.

- $\beta(\tau) \in \mathrm{End}_k(L)$: Let $\tau \in H^0(G, (S_n)_c)$ and $i \in \{1, \dots, m\}$ be arbitrary elements. We have to show that the image of $\tau(\iota_i)$ is contained in L_i. Equivalently, we can show that for all $s \in G_{L_i} := \mathrm{Gal}(K/L_i)$, we have $s \circ \tau(\iota_i) = \tau(\iota_i)$. So let s be an arbitrary element of G_{L_i}. By hypothesis, we know ${}^s\tau = \tau$, and using this we get:

$$\tau(\iota_i) = [{}^s\tau](\iota_i) \overset{(4.14)}{=} s \circ \tau(s^{-1}\iota_i) \overset{s^{-1}|_{L_i}=1}{=} s \circ \tau(\iota_i).$$

- $\beta(\tau)$ *injective:* Again, let τ be an arbitrary element of $H^0(G, (S_n)_c)$, and let $x = (x_1, \dots, x_m)$ be an element of the kernel of $\beta(\tau)$. We then have to show that all x_j are zero — so let $j \in \{1, \dots, m\}$ be an arbitrary index. The morphism $\tau^{-1}(\iota_j) : L \to K$ factorizes over L_i for a suitable i. Because G acts transitively on M, there is an $s \in G$ with $\tau^{-1}(\iota_j) = s\iota_i$. and because τ is fix under the G-action, so is τ^{-1}. Therefore we get

$$s\iota_i = \tau^{-1}(\iota_j) = [{}^s\tau^{-1}](\iota_j) = s \circ \tau^{-1}(s^{-1}\iota_j) \implies \tau(\iota_i) = s^{-1}\iota_j.$$

Applying both sides of this equation to x, we see:

$$0 = \tau(\iota_i)(x) = [s^{-1}\iota_j](x) = s^{-1}x_j,$$

and because s is an automorphism, it follows that x_j is zero.

So $\beta(\tau)$ is an injective endomorphism of the finite k-algebra L. But then it must also be *bijective* and consequently is an element of $\mathrm{Aut}_k(L)^{\mathrm{opp}}$ — which proves that β is a well defined *map*.

- $\beta\alpha = 1$: Let a be an arbitrary element of $\mathrm{Aut}_k(L)^{\mathrm{opp}}$. Then we have:

$$(\beta\alpha)(a) = \beta[M \to M, \varphi \mapsto \varphi a] = (\iota_1 a, \dots, \iota_m a)$$
$$= \underbrace{(\iota_1, \dots, \iota_m)}_{=1_L} \circ a = a.$$

- $\alpha\beta = 1$: Now let $\tau \in H^0(G, (S_n)_c)$ and $\varphi \in M$ be arbitrary elements. The morphism φ factorizes over L_i for a suitable i, and we find an $s \in G$ with $\varphi = s\iota_i$. Then we get:

$$[(\alpha\beta)(\tau)](\varphi) = \left[\alpha(\tau(\iota_1), \ldots, \tau(\iota_r))\right](\varphi) = \varphi \circ (\tau(\iota_1), \ldots, \tau(\iota_r))$$
$$= s \circ \tau(\iota_i) = s \circ \tau(s^{-1}s\iota_i) = [{}^s\tau](s\iota_i) = \tau(\varphi),$$

which concludes the proof of this lemma. (Note that even though we did not explicitly show that β is actually a homomorphism, it now follows automatically because α is an isomorphism and β is its inverse.) $\square$

Corollary 7.3 *Let $c \in Z^1_{cont}(G, S_n)$ be an arbitrary cocycle. Then, using the notation from lemma 4.17, we have the following commutative diagram of right-actions, where the right-action in the top row is the one associated to sequence (7.1) as defined in lemma 4.9:*

$$
\begin{array}{ccccc}
H^1_{cont}(G, (\mu_r^n)_c) & \times & H^0(G, (S_n)_c) & \longrightarrow & H^1_{cont}(G, (\mu_r^n)_c) \\
\gamma \uparrow \wr & & \alpha \uparrow \wr & & \gamma \uparrow \wr \\
L^\times / L^{\times r} & \times & \mathrm{Aut}_k(L)^{opp} & \longrightarrow & L^\times / L^{\times r} \\
(\bar{x} & , & a) & \longmapsto & \overline{a(x)}
\end{array}
$$

Proof. Let $x \in L^\times$ and $a \in \mathrm{Aut}_k(L)^{opp}$ be arbitrary elements. Then

$$\gamma(\overline{a(x)}) = \left(\left(\frac{s\sqrt[r]{s^{-1}\varphi a x}}{\sqrt[r]{\varphi a x}}\right)_{\varphi \in M}\right)_s \qquad \text{and}$$

$$\gamma(\bar{x})^{\alpha(a)} \overset{\text{lemma } 4.9}{=} \left(\alpha(a)^{-1} \cdot \left(\frac{s\sqrt[r]{s^{-1}\varphi x}}{\sqrt[r]{\varphi x}}\right)_{\varphi \in M} \cdot {}^s[\alpha(a)]\right)_s$$

$$\overset{(4.3)}{=} \left({}^{[\alpha(a)^{-1}]}\left(\frac{s\sqrt[r]{s^{-1}\varphi x}}{\sqrt[r]{\varphi x}}\right)_{\varphi \in M}\right)_s$$

$$= \left(\left(\frac{s\sqrt[r]{s^{-1}\varphi a x}}{\sqrt[r]{\varphi a x}}\right)_{\varphi \in M}\right)_s \cdot \qquad \square$$

Corollary 7.4   *Let $c \in Z^1_{cont}(G, S_n)$ be an arbitrary cocycle, and let us again use the notation from lemma 4.17. In particular, c is given by an fcs-algebra L of degree n over k: If we set $M := \operatorname{Hom}_k(L, K)$ and identify S_n with $\operatorname{Aut}(M)$, then $c_s = [M \to M,\ \varphi \mapsto s \circ \varphi]$.*

Let $p : \mu_r \wr S_n \twoheadrightarrow S_n$ be the canonical projection. Then we have the following bijection:

$$
\operatorname{Aut}_k(L) \backslash [L^\times / L^{\times^r}] \xrightarrow{\sim} \left\{ b \in H^1_{cont}(G, \mu_r \wr S_n) \;\middle|\; H^1_{cont}(p)(b) = [c] \right\}
$$

$$
\operatorname{Aut}_k(L) \cdot \bar{x} \quad \mapsto \quad \left(\left(\frac{s\sqrt[r]{s^{-1}\varphi x}}{\sqrt[r]{\varphi x}} \right)_{\varphi \in M} \cdot c_s \right)_s
$$

Here the left-action of $\operatorname{Aut}_k(L)$ on $L^\times / L^{\times^r}$ is the natural one, given by $^a\bar{x} := \overline{a(x)}$.

In particular, each class in $H^1_{cont}(G, \mu_r \wr S_n)$ is given by a pair (L, x) with an fcs-algebra L of degree n over k and an element x of $L^\times$. Two pairs (L, x) and (L', x') define the same class if and only if there is a k-isomorphism $\psi : L \xrightarrow{\sim} L'$, an element $y \in L^\times$ and an automorphism $a \in \operatorname{Aut}_k(L)$ such that $x' = \psi(a[xy^r])$.

Proof.   Corollary 4.6, applied to sequence (7.1), states

$$
H^1_{cont}(G, (\mu_r^n)_c) / H^0(G, (K^\times)^n_c) \xrightarrow{\sim} H^1_{cont}(p)^{-1}[c] \subseteq H^1_{cont}(G, \mu_r \wr S_n)
$$

$$
(a_s) \cdot H^0(G, (K^\times)^n_c) \quad \mapsto \quad (a_s \cdot c_s).
$$

Using corollary 7.1 and lemma 7.6, we thus get the following commutative diagram:

$$
\begin{array}{ccc}
H^1_{cont}(G, (\mu_r^n)_c) / H^0(G, (K^\times)^n_c) & \xrightarrow{\ \sim\ } & H^1_{cont}(p)^{-1}[c] \\
\big\uparrow{\scriptstyle \wr} & \nearrow{\scriptstyle \sim} & \\
[L^\times / L^{\times^r}] / \operatorname{Aut}_k(L)^{\mathrm{opp}} & &
\end{array}
$$

Here we can replace $[L^\times / L^{\times^r}] / \operatorname{Aut}_k(L)^{\mathrm{opp}}$ by $\operatorname{Aut}_k(L) \backslash [L^\times / L^{\times^r}]$, because the right-action of a group is the same as a left-action of the opposite group. Corollary 7.3 shows that the action of $\operatorname{Aut}_k(L)$ on $L^\times / L^{\times^r}$ is the natural one, and lemma 7.1 shows that the resulting bijection from

$\mathrm{Aut}_k(L)\backslash[L^\times/L^{\times^r}]$ to $H^1_{\mathrm{cont}}(p)^{-1}[c]$ is given by the formula stated in the corollary. $\qquad\square$

Proposition 7.2 *Consider the special case where $k = \mathbb{F}_q$ is a finite field, and let $b \in H^1_{\mathrm{cont}}(G, \mu_r \wr S_n)$ be an arbitrary class, given by a pair (L, x) according to corollary 7.4.*

Let $L = \prod_{i=1}^m L_i$ with subfields L_i of $K = \overline{\mathbb{F}}_q$ of degree n_i over $\mathbb{F}_q$, and let $x = (x_1, \ldots, x_m)$ with $x_i \in L_i^\times$. For each i, choose an r-th root y_i of x_i in K. According to lemma 4.3, an 1-cocycle (b_s) representing b is uniquely determined by b_f, where f is the Frobenius automorphism $y \mapsto y^q$ of G. In this sense, b is given by

$$\boxed{\; b_f = \prod_{i=1}^m \left[\left(y_i^{q^{n_i}-1}, 1, \ldots, 1 \right) \cdot z_i \right] \in \mu_r \wr S_n \;}$$

Here z_i denotes the cycle $[1, \ldots, n_i]$ in $S_{n_i} \le S_n$.

Proof. Because of lemma 4.18 and lemma 4.5, we only have to consider the case $m = 1$, i.e. L is a subfield of K of degree n over $\mathbb{F}_q$, and $y = y_1$ is an r-th root of $x \in L^\times$.

It is $M := \mathrm{Hom}_k(L, K) = \{f^0, \ldots, f^{n-1}\}$, and we want to identify M with the set $\{0, \ldots, n-1\}$ via $f^i \mapsto i$. For all $i \in \{0, \ldots, n-1\}$, we can choose $y^{(q^i)}$ as an r-th root of $f^i x = x^{(q^i)}$, and because of corollary 7.4, in order to prove the corollary, we have to show that with these choices of r-th roots

$$\frac{f\sqrt[r]{f^{-1}\varphi x}}{\sqrt[r]{\varphi x}} = (y^{q^{n-1}-1}, 1, \ldots, 1)$$

and that $c_f = [1, \ldots, n]$. The latter is obvious since $f \circ f^i = f^{i+1}$.

First consider the case $i = 0$. We get:

$$\frac{f\sqrt[r]{f^{-1}f^0 x}}{\sqrt[r]{f^0 x}} = \frac{f\sqrt[r]{f^{-1}x}}{\sqrt[r]{x}} = \frac{f\sqrt[r]{f^{n-1}x}}{y} = \frac{f y^{(q^{n-1})}}{y}$$

$$= \frac{\left(y^{(q^{n-1})} \right)^q}{y} = \frac{y^{(q^{n-1+1})}}{y} = y^{q^n-1}.$$

Now look at the case $i \geq 1$:

$$\frac{f\sqrt[r]{f^{-1}f^i x}}{\sqrt[r]{f^i x}} = \frac{f\sqrt[r]{f^{i-1}x}}{\sqrt[r]{f^i x}} = \frac{fy^{(q^{i-1})}}{y^{(q^i)}} = \frac{y^{(q^i-1+1)}}{y^{(q^i)}} = \frac{y^{(q^i)}}{y^{(q^i)}} = 1.$$

$\square$

Example 7.1   Consider the special case $n = 6$, $r = 3$ and $k = \mathbb{F}_7$ (and thus $K = \overline{\mathbb{F}}_7$).

It is $\mathbb{F}_{49} = \mathbb{F}_7(\alpha)$ with $\alpha^2 + 5\alpha + 5 = 0$ and $\mathbb{F}_{2401} = \mathbb{F}_7(\beta)$ with $\beta^4 + 5\beta^3 + 4\beta^2 + \beta + 5 = 0$; with this, define the class $b \in H^1_{\text{cont}}(G, \mu_3 \wr S_6)$ as the pair $\left(\mathbb{F}_{2401} \times \mathbb{F}_{49}, \left(\frac{1}{\beta}, \frac{1}{\alpha^2}\right)\right)$.

So we have $n_1 = 4$ and $n_2 = 2$, choose third roots y_1 of $\frac{1}{\beta}$ and y_2 of $\frac{1}{\alpha^2}$ and calculate:

$$y_1^{7^4-1} = \left(\frac{1}{\beta}\right)^{\frac{7^4-1}{3}} = \left(\frac{1}{\beta}\right)^{800} = \beta^{2400-800} = \beta^{1600} = \left(\beta^{400}\right)^4$$
$$= \left(\beta^{1+7+49+343}\right)^4 = \left[N_{\mathbb{F}_{2401}/\mathbb{F}_7}(\beta)\right]^4 = 5^4 = 2,$$

$$y_2^{7^2-1} = \left(\frac{1}{\alpha^2}\right)^{\frac{7^2-1}{3}} = \left(\frac{1}{\alpha^2}\right)^{16} = \alpha^{48-2\cdot16} = \alpha^{16} = \left(\alpha^8\right)^2$$
$$= \left(\alpha^{1+7}\right)^2 = \left[N_{\mathbb{F}_{49}/\mathbb{F}_7}(\alpha)\right]^2 = 5^2 = 4.$$

Now proposition 7.2 implies that an 1-cocycle (b_s) representing the class b is given by

$$b_f = (2,1,1,1,4,1) \cdot [1234][56] \in \mu_3 \wr S_6.$$

Corollary 7.5   *Again, let $k = \mathbb{F}_q$ be a finite field, and let $L = \prod_{i=1}^m \mathbb{F}_{q^{n_i}}$ be an fcs-algebra of degree n over k. Choose an $N \in \mathbb{N}$ sufficiently large such that $\mathbb{F}_{q^N}$ contains all the $\mathbb{F}_{q^{n_i}}$ as subfields as well as all r-th roots of all elements of all the $\mathbb{F}_{q^{n_i}}$. Let $\alpha \in \mathbb{F}_{q^N}^\times$ be a generator of the multiplicative group $\mathbb{F}_{q^N}^\times$.*

Then for each $i \in \{1, \ldots, m\}$, the element $\alpha_i := \alpha^{\frac{q^N-1}{q^{n_i}-1}}$ is a generator of $\mathbb{F}_{q^{n_i}}^\times$, and $\zeta := \alpha^{\frac{q^N-1}{r}}$ is a primitive r-th root of unity.

Now let $x = (\alpha_1^{k_1}, \ldots, \alpha_m^{k_m}) \in L^\times$ for natural numbers $k_1, \ldots, k_m$. Then the class in $H^1_{\text{cont}}(G, \mu_r \wr S_n)$ defined by the pair (L, x) in the sense of corollary

7.4 is represented by an 1-cocycle (b_s) which on the Frobenius $f : \overline{\mathbb{F}}_q \to \overline{\mathbb{F}}_q$, $y \mapsto y^q$, is given by

$$b_f = \prod_{i=1}^{m} \left[(\zeta^{k_i}, 1, \ldots, 1) \cdot z_i \right]$$

Proof. The index of $\mathbb{F}_{q^{n_i}}^{\times}$ in $\mathbb{F}_{q^N}^{\times}$ is obviously $\frac{q^N-1}{q^{n_i}-1}$, which shows that the α_i are indeed generators of $\mathbb{F}_{q^{n_i}}^{\times}$. It is also clear that μ_r, which by hypothesis is contained in $\mathbb{F}_{q^N}^{\times}$, is generated by ζ, because the index of μ_r in $\mathbb{F}_{q^N}^{\times}$ is $\frac{q^N-1}{r}$. By choice of N, for each i the r-th roots of α_i are contained in $\mathbb{F}_{q^N}$, and one of these roots is clearly given by $\alpha^{\frac{q^N-1}{r \cdot q^{n_i}-1}}$. Thus $y_i := \alpha^{\frac{k_i \cdot (q^N-1)}{r \cdot (q^{n_i}-1)}} \in \mathbb{F}_{q^N}$ is an r-th roots of $\alpha_i^{k_i}$. So according to proposition 7.2, we only have to show that $y_i^{q^{n_i}-1}$ equals ζ^{k_i}:

$$y_i^{q^{n_i}-1} = \alpha^{\frac{k_i \cdot (q^N-1)}{r \cdot (q^{n_i}-1)} \cdot (q^{n_i}-1)} = \left(\alpha^{\frac{q^N-1}{r}} \right)^{k_i} = \zeta^{k_i}.$$

$\square$

Lemma 7.7

(1) If L/k is a separable field extension and if $\nu \in \mathbb{N}_+$ is a natural number, then

$$\mathrm{Aut}_k(L^{\nu}) = \mathrm{Aut}_k(L) \wr S_{\nu},$$

where the wreath product acts on L^{ν} as follows:

$$[(a_i)_i \cdot \sigma](x_j)_j = \left(a_j x_{\sigma^{-1}j} \right)_j .$$

(2) If $L = \prod_{i=1}^{m} L_i^{\nu_i}$ is an fcs-algebra over k with separable field extensions L_i of k of which no two are isomorphic, then

$$\mathrm{Aut}_k(L) = \prod_{i=1}^{m} \mathrm{Aut}_k(L_i^{\nu_i}) \overset{(1)}{=} \prod_{i=1}^{m} \left[\mathrm{Aut}_k(L_i) \wr S_{\nu_i} \right].$$

Proof. Consider the map

$$\varphi : \mathrm{Aut}_k(L) \wr S_{\nu} \longrightarrow \mathrm{End}_k(L^{\nu})$$
$$(a_i)_i \cdot \sigma \quad \mapsto \quad \left[(x_j)_j \mapsto (a_j x_{\sigma^{-1}j})_j \right]$$

which is obviously well defined. We first claim that it is a homomorphism (of monoids): It is clear that $(1_L)_i \cdot 1_{s_\nu}$ is mapped to 1_{L^ν}, and for elements $(a_i)_i \cdot \sigma$ and $(b_i)_i \cdot \tau$ of $\mathrm{Aut}_k(L) \wr S_\nu$ and $x = (x_j)_j$ of L^ν we have

$$\varphi\Big([(a_i)_i \cdot \sigma] \cdot [(b_i)_i \cdot \tau] \Big)(x) = \varphi\Big((a_i b_{\sigma^{-1}i})_i \cdot \sigma\tau \Big)(x)$$
$$= \Big([a_j b_{\sigma^{-1}j}] x_{(\sigma\tau)^{-1}j} \Big)_j = \Big([a_j b_{\sigma^{-1}j}] x_{\tau^{-1}\sigma^{-1}j} \Big)_j$$
$$= \Big(a_j [b_{\sigma^{-1}j} x_{\tau^{-1}\sigma^{-1}j}] \Big)_j = \varphi[(a_i)_i \cdot \sigma]\Big((b_j x_{\tau^{-1}j})_j \Big)$$
$$= \varphi[(a_i)_i \cdot \sigma]\Big(\varphi[(b_i)_i \cdot \tau](x) \Big).$$

Thus φ induces a homomorphism of groups

$$\varphi : \mathrm{Aut}_k(L) \wr S_\nu \longrightarrow \mathrm{Aut}_k(L^\nu)$$

which is clearly injective. To prove surjectivity, let a be a k-automorphism of L^ν, and for $i \in \{1, \ldots, \nu\}$ denote the i-th factor of L^ν by $L^{(i)}$ and the composition $L^\nu \xrightarrow{a} L^\nu \to L^{(i)}$ by $a^{(i)}$. Because L is a field, $a^{(i)}$ must factor over one of the projections $L^\nu \to L^{(j)}$, say over the $\tau(i)$-th projection. In this way we get a map $\sigma : \{1, \ldots, \nu\} \to \{1, \ldots, \nu\}$, and for each i a commutative diagram

$$
\begin{array}{ccc}
L^\nu & \xrightarrow{\ \ a\ \ } & L^\nu \\
{\scriptstyle \mathrm{pr}_{\tau i}}\Big\downarrow & \sim & \Big\downarrow{\scriptstyle \mathrm{pr}_i} \\
L^{(\tau i)} & \xdashrightarrow{\ a_i\ } & L^{(i)}.
\end{array}
$$

This means that for $x = (x_j)_j$ we have $a(x) = (a_j x_{\tau j})_j$ which easily implies that all a_i are automorphisms of L.

Furthermore, we claim that τ is a permutation: By construction of τ, the automorphism a factors over $\prod_{j \in \tau(\{1,\ldots,\nu\})} L^{(j)}$, so a could not be injective if $\tau(\{1, \ldots, \nu\})$ was not the whole of $\{1, \ldots, \nu\}$. This shows that τ is surjective and hence that it is a permutation. This completes the proof of (1), because we clearly have $a = \varphi[(a_i) \cdot \tau^{-1}]$.

To prove (2), consider the map

$$\psi : \prod_{i=1}^m \mathrm{Aut}_k(L_i^{\nu_i}) \longrightarrow \mathrm{Aut}_k(L)$$
$$(a_i)_i \quad \mapsto \quad [(x_j)_j \mapsto (a_j x_j)_j]$$

which is clearly a well defined, injective group homomorphism. To show that ψ is surjective, let a be a k-automorphism of L. Arguing exactly as above, we get commutative diagrams

$$
\begin{array}{ccc}
L & \xrightarrow[\sim]{a} & L \\
{\scriptstyle\mathrm{pr}_{\tau i}}\downarrow & & \downarrow{\scriptstyle\mathrm{pr}_i} \\
L_{\tau i}^{\nu_{\tau i}} & \xdashrightarrow[a_i]{\sim} & L_i^{\nu_i}
\end{array}
$$

for each $i \in \{1, \ldots, m\}$. Now we can repeat the same reasoning for each i to get commutative diagrams

$$
\begin{array}{ccc}
L_{\tau i}^{\nu_{\tau i}} & \xrightarrow[\sim]{a_i} & L_i^{\nu_i} \\
{\scriptstyle\mathrm{pr}_{\tau_{ij}}}\downarrow & & \downarrow{\scriptstyle\mathrm{pr}_i} \\
L_{\tau i}^{(\tau_i j)} & \xdashrightarrow[a_j^{(i)}]{\sim} & L_i^{(j)}
\end{array}
$$

for each $j \in \{1, \ldots, \nu_i\}$ — here τ_j is a map from $\{1, \ldots, \nu_i\}$ to $\{1, \ldots, \nu_{\tau i}\}$. Thus $L_{\tau i}$ is isomorphic to L_i which by hypothesis is only possible if $\tau i = i$. This shows that all a_i are automorphisms of $L_i^{\nu_i}$ and that $a = \psi(a_1, \ldots, a_m)$, which completes the proof of the lemma. $\qquad\square$

Example 7.2 In this example, we want to show how corollary 7.4 and lemma 7.7 allow us to efficiently compute the cohomology $H_{\mathrm{cont}}^1(G, \mu_r \wr S_n)$ for given k, n and r.

(1) Let $k := \mathbb{R}$ (and thus $K = \mathbb{C}$), $n = 3$ and $r = 4$. There are — up to $\mathbb{R}$-isomorphism — exactly two fcs-algebras of degree three over $\mathbb{R}$, namely $\mathbb{R}^3$ and $\mathbb{C} \times \mathbb{R}$.

First let c be the class defined by $L = \mathbb{R}^3$ in $H_{\mathrm{cont}}^1(G, S_3)$ (which of course is the trivial class). We have $\mathbb{R}^\times/(\mathbb{R}^\times)^4 = \{\overline{-1}, \overline{1}\}$ and therefore $L^\times/L^{\times 4} = \{\overline{-1}, \overline{1}\}^3$. According to lemma 7.7(1), it is $\mathrm{Aut}_{\mathbb{R}}(L) = \mathrm{Aut}_{\mathbb{R}}(\mathbb{R}) \wr S_3 = S_3$, and under the action of this group $L^\times/L^{\times 4}$ decomposes into four orbits which are represented by $(\overline{1}, \overline{1}, \overline{1})$, $(\overline{1}, \overline{1}, \overline{-1})$, $(\overline{1}, \overline{-1}, \overline{-1})$ and $(\overline{-1}, \overline{-1}, \overline{-1})$.

Now let c be the class defined by $L = \mathbb{C} \times \mathbb{R}$ in $H_{\mathrm{cont}}^1(G, S_3)$. Since

$\mathbb{C}^\times = \mathbb{C}^{\times 4}$, we have $L^\times/L^{\times 4} = \{(\bar{1},\bar{1}),(\bar{1},\overline{-1})\}$. Further

$$\mathrm{Aut}_{\mathbb{R}}(L) \stackrel{\text{lemma } 7.7(2)}{=} \mathrm{Aut}_{\mathbb{R}}(\mathbb{C}) \times \mathrm{Aut}_{\mathbb{R}}(\mathbb{R}) = \mathrm{Aut}_{\mathbb{R}}(\mathbb{C}) = \{1,\tau\}$$

with $\tau(a,b) = (\bar{a},b)$, and this group obviously acts trivially on $L^\times/L^{\times 4}$, so that in this case we get two orbits.

So corollary 7.4 shows that there are six classes in $H^1_{\mathrm{cont}}(G,\mu_4 \wr S_3)$, given by

$$\bigl(\mathbb{R}^3,(1,1,1)\bigr), \bigl(\mathbb{R}^3,(1,1,-1)\bigr), \bigl(\mathbb{R}^3,(1,-1,-1)\bigr), \bigl(\mathbb{R}^3,(-1,-1,-1)\bigr),$$
$$\bigl(\mathbb{C}\times\mathbb{R},(1,1)\bigr) \text{ and } \bigl(\mathbb{C}\times\mathbb{R},(1,-1)\bigr).$$

(2) Now let $k := \mathbb{F}_5$, $n = 4$ and $r = 3$. Then there are — up to k-isomorphism — the following fcs-algebras of degree four over $\mathbb{F}_5$: $\mathbb{F}_5^4$, $\mathbb{F}_{25} \times \mathbb{F}_5^2$, $\mathbb{F}_{25}^2$, $\mathbb{F}_{125} \times \mathbb{F}_5$ and $\mathbb{F}_{625}$.

Using lemma 7.7 and denoting the Frobenius $a \mapsto a^5$ by f, we get:

$$\mathrm{Aut}_{\mathbb{F}_5}(\mathbb{F}_5^4) \quad = \mathrm{Aut}_{\mathbb{F}_5}(\mathbb{F}_5) \wr S_4 \qquad\qquad = S_4,$$

$$\mathrm{Aut}_{\mathbb{F}_5}(\mathbb{F}_{25} \times \mathbb{F}_5^2) = \mathrm{Aut}_{\mathbb{F}_5}(\mathbb{F}_{25}) \times \bigl[\mathrm{Aut}_{\mathbb{F}_5}(\mathbb{F}_5) \wr S_2\bigr]$$
$$= \{1,f\} \times S_2,$$

$$\mathrm{Aut}_{\mathbb{F}_5}(\mathbb{F}_{25}^2) \quad = \mathrm{Aut}_{\mathbb{F}_5}(\mathbb{F}_{25}) \wr S_2 \qquad = \{1,f\} \wr S_2,$$

$$\mathrm{Aut}_{\mathbb{F}_5}(\mathbb{F}_{125} \times \mathbb{F}_5) = \mathrm{Aut}_{\mathbb{F}_5}(\mathbb{F}_{125}) \times \mathrm{Aut}_{\mathbb{F}_5}(\mathbb{F}_5)$$
$$= \{1 \times 1, f \times 1, f^2 \times 1\},$$

$$\mathrm{Aut}_{\mathbb{F}_5}(\mathbb{F}_{625}) \quad = \qquad\qquad\qquad \{1,f,f^2,f^3\}.$$

First let $L := \mathbb{F}_5^4$. Since $(\#\mathbb{F}_5^\times, 3) = (4,3) = 1$, taking the third power is bijective in $\mathbb{F}_5^\times$. Consequently, already $\mathbb{F}_5^\times/\mathbb{F}_5^{\times 3}$ is a singleton set, so that $L^\times/L^{\times 3}$ and $\mathrm{Aut}_{\mathbb{F}_5}(L)\backslash L^\times/L^{\times 3}$ of course are singleton sets as well. Thus only the trivial class of $H^1_{\mathrm{cont}}(G,\mu_3 \wr S_4)$ is mapped to the class represented by L in $H^1_{\mathrm{cont}}(G,S_4)$.

Now let $L := \mathbb{F}_{25} \times \mathbb{F}_5^2$. Since $3 \mid \#\mathbb{F}_{25}^\times = 24$, taking the third power is *not* bijective in $\mathbb{F}_{25}^\times$. On the contrary, the third powers constitute a subgroup of index three in $\mathbb{F}_{25}^\times$. Let $\alpha \in \mathbb{F}_{25}^\times$ be a generator of $\mathbb{F}_{25}^\times$ (for example, we can take a root of the polynomial $T^2 + 2T + 3 \in \mathbb{F}_5[T]$). Then $\mathbb{F}_{25}^\times/\mathbb{F}_{25}^{\times 3} = \{\bar{1},\bar{\alpha},\overline{\alpha^2}\}$. Because $f(\alpha) = \alpha^5 = \alpha^3 \cdot \alpha^2$, the elements $(\bar{\alpha},\bar{1},\bar{1})$ and $(\overline{\alpha^2},\bar{1},\bar{1})$ of $L^\times/L^{\times 3}$ are in the same $(\{1,f\} \times S_2)$-orbit, and it follows that for this choice of L, there are exactly two cohomology classes, represented by $(1,1,1)$ and $(\alpha,1,1)$.

In the case $L := \mathbb{F}_{25}^2$, we have $\mathrm{Aut}_{\mathbb{F}_5}(L) = \{1,f\} \wr S_2$, and what we know about $\mathbb{F}_{25}^\times / \mathbb{F}_{25}^{\times 3}$ implies $\mathrm{Aut}_{\mathbb{F}_5}(L) \backslash (L^\times / L^{\times 3}) = \{(\bar{1},\bar{1}), (\bar{1},\bar{\alpha}), (\bar{\alpha},\bar{\alpha})\}$.

For $L := \mathbb{F}_{125} \times \mathbb{F}_5$ we have $(\mathbb{F}_{125}^\times, 3) = (124, 3) = 1$, so as in the first case it follows that $L^\times = L^{\times 3}$ and that there is only one cohomology class, represented by $(1,1)$.

Finally, let $L := \mathbb{F}_{625}$, and let γ denote a root of the polynomial $T^4 + T^2 + 3T + 3 \in \mathbb{F}_5[T]$. Then γ is a generator of $\mathbb{F}_{625}^\times$, and $\gamma^{26} = \gamma^3 + 2\gamma^2 + 2\gamma + 1$ is a root of the polynomial $T^2 + 2T + 3$, which means we can put $\alpha := \gamma^{26}$. Like before we see that $L^\times / L^{\times 3} = \{\bar{1}, \bar{\gamma}, \overline{\gamma^2}\}$ and that $\bar{\gamma}$ and $\overline{\gamma^2}$ are in the same $\mathrm{Aut}_{\mathbb{F}_5}(L)$-orbit — so again we have two cohomology classes, namely those represented by 1 and γ.

We conclude that there are the following nine classes in $H^1_{\mathrm{cont}}(G, \mu_3 \wr S_4)$:

$$\left(\mathbb{F}_5^4, (1,1,1,1)\right),$$
$$\left(\mathbb{F}_{25} \times \mathbb{F}_5^2, (1,1,1)\right), \left(\mathbb{F}_{25} \times \mathbb{F}_5^2, (\alpha,1,1)\right),$$
$$\left(\mathbb{F}_{25}^2, (1,1)\right), \left(\mathbb{F}_{25}^2, (1,\alpha)\right), \left(\mathbb{F}_{25}^2, (\alpha,\alpha)\right),$$
$$\left(\mathbb{F}_{125} \times \mathbb{F}_5, (1,1)\right),$$
$$\left(\mathbb{F}_{625}, 1\right) \text{ and } \left(\mathbb{F}_{625}, \gamma\right).$$

Proposition 7.3 *Let $b \in H^1_{cont}(G, \mu_r \wr S_n)$ be an arbitrary cohomology class, given by a pair (L,x) according to corollary 7.4. Like always let $L = \prod_{i=1}^m L_i$ with subfields L_i of K of degree n_i over k, and let $x = (x_1, \ldots, x_m)$ with $x_i \in L_i^\times$.*

For each $i \in \{1, \ldots, m\}$, choose a k-basis $e_1^{(i)}, \ldots, e_{n_i}^{(i)}$ of L_i, and let $M_i := \mathrm{Hom}_k(L_i, K)$. With this define

$$P_n^r\{b\} := \sum_{i=1}^m Tr_{L_i/k} \left[\frac{1}{x_i} \cdot \left(\sum_{j=1}^{n_i} e_j^{(i)} X_j^{(i)} \right)^r \right] \in k[X_j^{(i)}], \qquad (7.3)$$

where $Tr_{L_i/k} : L_i[X_j^{(i)}] \to k[X_j^{(i)}]$ is the k-linear map sending monomials $X_1^{k_1} \cdot \ldots \cdot X_n^{k_n}$ to themselves and constants in L_i to their trace in k. Then $P_n^r\{b\}$, the Fermat equation twisted by b, is an $\mathcal{F}_K^{n,r} / \mathcal{F}_k^{n,r}$-form of P_n^r, and we have $\vartheta(P_n^r\{b\}) = b$. *In particular, the k-isomorphism class of $P_n^r\{b\}$ only depends on b and not on any choices made.*

Proof. Let i be an arbitrary element of $\{1,\dots,m\}$. For $\varphi \in M_i$, consider the morphism of k-algebras from $L_i[X_j^{(i)}]$ to $K[X_j^{(i)}]$ which is given by φ on L and which sends $X_j^{(i)}$ to itself, and call this morphism φ as well. Then it is obviously true that $\sum_{\varphi \in M_i} \varphi = \mathrm{Tr}_{L_i/k} : L_i[X_j^{(i)}] \to K[X_j^{(i)}]$. Now define the matrix

$$B_i := \left(\frac{\varphi e_j^{(i)}}{\sqrt[r]{\varphi x_i}} \right)_{\varphi \in M_i,\, j \in \{1,\dots,n_i\}} \in M(n_i \times n_i, K).$$

Then we have

$$\det(B_i) = \underbrace{\frac{1}{\prod_{\varphi \in M_i} \sqrt[r]{\varphi x_i}}}_{\neq 0} \cdot \underbrace{\det\left(\varphi e_j^{(i)} \right)_{\varphi,j}}_{\neq 0} \neq 0,^2$$

i.e. B_i is an element of $\mathrm{GL}\,(n_i, K)$. Let B be the block matrix built from the B_i for $i = 1,\dots,m$ — so B is an invertible $n \times n$-matrix over K.

Now we are going to show that $P_n^r(BX) = P_n^r\{b\}$, so that B defines an isomorphism from $P_n^r\{b\}$ to P_n^r:

$$P_n^r(BX) = \sum_{i=1}^m \sum_{\varphi \in M_i} \left(\sum_{j=1}^{n_i} \frac{\varphi e_j^{(i)}}{\sqrt[r]{\varphi x_i}} \cdot X_j^{(i)} \right)^r$$

$$= \sum_{i=1}^m \sum_{\varphi \in M_i} \left(\frac{1}{\varphi x_i} \right) \cdot \left(\sum_{j=1}^{n_i} \varphi e_j^{(i)} \cdot X_j^{(i)} \right)^r$$

$$= \sum_{i=1}^m \sum_{\varphi \in M_i} \varphi \left[\frac{1}{x_i} \cdot \left(\sum_{j=1}^{n_i} e_j^{(i)} \cdot X_j^{(i)} \right)^r \right]$$

$$= \sum_{i=1}^m \mathrm{Tr}_{L_i/k} \left[\frac{1}{x_i} \cdot \left(\sum_{j=1}^{n_i} e_j^{(i)} \cdot X_j^{(i)} \right)^r \right]$$

---

[2] According to [Lan93, 5.4, p286], the following is true: If L/k is an arbitrary separable field extension of degree n, if $\{e_1,\dots,e_n\}$ is a k-basis of L, and if $\varphi_1,\dots,\varphi_n : L \to K$ are the different k-embeddings of L into a separable algebraic closure K of L, then the matrix $(\varphi_i e_j)_{i,j}$ is invertible.

$$= P_n^r\{b\}.$$

So $P_n^r\{b\}$ is indeed an $\mathcal{F}_K^{n,r}/\mathcal{F}_k^{n,r}$-form of P_n^r, and we have $\vartheta[P_n^r\{b\}] = [(C_s)_s]$ with $C_s := B^s(B^{-1})$.

Let (b_s) be the 1-cocycle of G in $\mathrm{GL}\,(n, K)$ defined by (L, x), which cohomology class of course is b. Using lemma 7.1 and corollary 7.4, we get the following formula for b_s (where we want to choose the same r-th roots that we have chosen in the definition of B):

$$b_s = \left(\frac{s\sqrt[r]{s^{-1}\varphi x}}{\sqrt[r]{\varphi x}} \cdot \delta_{\varphi, s\psi} \right)_{\varphi, \psi \in M}$$

So like B, the matrix b_s is a block matrix with blocks $b_s^{(i)}$ (for $i = 1, \ldots, m$) given by

$$b_s^{(i)} = \left(\frac{s\sqrt[r]{s^{-1}\varphi x_i}}{\sqrt[r]{\varphi x_i}} \cdot \delta_{\varphi, s\psi} \right)_{\varphi, \psi \in M_i}.$$

We are done if we can show that for arbitrary $s \in G$, we have $C_s = b_s$ or equivalently $B = b_s{}^s B$ which means $B_i = b_s^{(i)}{}^s B_i$ for all i:

$$
\begin{aligned}
b_s^{(i)}{}^s B_i &= \left(\frac{s\sqrt[r]{s^{-1}\varphi x_i}}{\sqrt[r]{\varphi x_i}} \cdot \delta_{\varphi, s\psi} \right)_{\varphi, \psi \in M_i} \cdot \left(\frac{s\psi e_j^{(i)}}{s\sqrt[r]{\psi x_i}} \right)_{\psi \in M_i, j \in \{1,\ldots,n_i\}} \\
&= \left(\sum_{\psi \in M_i} \left[\frac{s\sqrt[r]{s^{-1}\varphi x_i}}{\sqrt[r]{\varphi x_i}} \cdot \delta_{\varphi, s\psi} \cdot \frac{s\psi e_j^{(i)}}{s\sqrt[r]{\psi x_i}} \right] \right)_{\varphi \in M_i, j \in \{1,\ldots,n_i\}} \\
&= \left(\frac{s\sqrt[r]{s^{-1}\varphi x_i}}{\sqrt[r]{\varphi x_i}} \cdot \frac{s(s^{-1}\varphi)e_j^{(i)}}{s\sqrt[r]{(s^{-1}\varphi)x_i}} \right)_{\varphi \in M_i, j \in \{1,\ldots,n_i\}} \\
&= \left(\frac{\varphi e_j^{(i)}}{\sqrt[r]{\varphi x_i}} \right)_{\varphi \in M_i, j \in \{1,\ldots,n_i\}} \\
&= B_i.
\end{aligned}
$$

$\square$

Corollary 7.6 *For $r \geq 3$, the map ϑ is not only injective, but also surjective, so that we get a canonical isomorphism*

$$\boxed{\vartheta : E\left(\mathcal{F}_K^{n,r}/\mathcal{F}_k^{n,r}, P_n^r\right) \xrightarrow{\sim} H_{cont}^1\left(G, \mu_r \wr S_n\right)}$$

of pointed sets.

Proof.   Theorem 6.1 states that the canonical map

$$\vartheta : E\left(\mathcal{F}_K^{n,r}/\mathcal{F}_k^{n,r}, P_n^r\right) \longrightarrow H_{\text{cont}}^1\left(G, A(P_n^r)\right)$$

is *injective*, and because of proposition 7.1 and lemma 7.2 we can iden-
tify $H_{\text{cont}}^1\left(G, A(P_n^r)\right)$ with $H_{\text{cont}}^1\left(G, \mu_r \wr S_n\right)$ since by hypothesis $r \geq 3$.
Finally, the map ϑ is *surjective* as well, because for a given class b in
$H_{\text{cont}}^1\left(G, \mu_r \wr S_n\right)$, the k-isomorphism class of the twisted Fermat equation
$P_n^r\{b\}$ is a preimage of b under ϑ according to proposition 7.3. $\square$

Remark 7.1   *Let A be an arbitrary commutative ring with unit, and let B
be a projective A-algebra of finite rank. As shown for example in [Brü98],
in this situation one can define a canonical A-linear map $Tr_{B/A} : B \longrightarrow A$,
the* trace *from B to A, which has the following properties:*

(1) If B is a free A-algebra, then $Tr_{B/A}$ is the usual trace map.
(2) If A' is an arbitrary A-algebra, then $Tr_{(B \otimes_A A')/A'} = Tr_{B/A} \otimes 1_{A'}$.
*(3) If $B = \prod_{i=1}^m B_i$ is a finite product of projective A-algebras B_i of finite
rank, then $Tr_{B/A} = \sum_i Tr_{B_i/A}$.*
*(4) If C is a projective B-algebra of finite rank, then $Tr_{C/A} = Tr_{B/A} \circ
Tr_{C/B}$.*

*Using the notation from proposition 7.3 and properties (1)-(3) of the trace,
we get:*

$$\boxed{P_n^r\{b\} = Tr_{L[X_j^{(i)}]/k[X_j^{(i)}]} \left[\frac{1}{x} \left(\sum_{i,j} e_j^{(i)} X_j^{(i)} \right)^r \right].}$$

Corollary 7.7   *Again, let $r \geq 3$, and consider an arbitrary field extension
k'/k and a separable algebraic closure K' of k'. For a polynomial $Q \in
k[X_1, \ldots, X_n]$, denote the image of Q under $k[X_i] \overset{can}{\hookrightarrow} k'[X_i]$ by Q as well.*

Then we have the following commutative diagram of pointed sets:

$$
\begin{array}{ccccccc}
[Q] & \in & E(\mathcal{F}_K^{n,r}/\mathcal{F}_k^{n,r}, P_n^r) & \xrightarrow[\sim]{\vartheta} & H_{cont}^1(G_k, Aut_K(P_n^r)) & \ni & (L,x) \\
\downarrow & & \downarrow & & \downarrow & & \downarrow \\
[Q] & \in & E(\mathcal{F}_{K'}^{n,r}/\mathcal{F}_{k'}^{n,r}, P_n^r) & \xrightarrow[\vartheta]{\sim} & H_{cont}^1(G_{k'}, Aut_{K'}(P_n^r)) & \ni & (L\otimes_k k', x\otimes 1)
\end{array}
$$

Here we again identify cohomology classes with pairs (L, x) in the sense of corollary 7.4.

Proof. First we want to check that the vertical maps are well defined: We can embed K into K' and then assume without loss of generality that $K \subseteq K'$. If there is an isomorphism from Q to P_n^r defined over K, then this isomorphism is in particular defined over K', and if Q and P_n^r are already isomorphic over k, then they are in particular isomorphic over k' — this shows that the left vertical map is well defined.

If L is an fcs-algebra of degree n over k, then $L \otimes_k k'$ obviously is an fcs-algebra of degree n over k', and if x is invertible in L, then $x \otimes 1$ is invertible in $L \otimes_k k'$. Furthermore, if (L, x) and (L', x') define the same cohomology class, then according to corollary 7.4 there is a k-isomorphism $\psi : L \xrightarrow{\sim} L'$, an element $y \in L^\times$ and an automorphism $a \in Aut_k(L)$ with $x' = \psi(a[xy^r])$. But then $\psi \otimes 1_{k'}$ is a k'-isomorphism from $L \otimes_k k'$ to $L' \otimes_k k'$, and we have $x' \otimes 1 = (\psi \otimes 1_{k'})((a \otimes 1_{k'})[(x \otimes 1)(y \otimes 1)^r])$, which shows that $(L \otimes_k k', x \otimes 1)$ and $(L' \otimes_k k', x' \otimes 1)$ represent the same cohomology class. — Therefore also the right vertical map is well defined, and it is obvious that both vertical maps are actually morphisms of pointed sets.

Because ϑ is an isomorphism according to corollary 7.6, we can prove that the diagram commutes by taking a pair (L, x) representing an element in the upper right corner and showing that it is mapped to the same element in the lower left corner under both ways through the diagram:

Under ϑ^{-1}, the pair (L, x) is mapped to $[P_n^r\{(L, x)\}]$ because of proposition 7.3. We therefore have to show that $P_n^r\{(L, x)\}$ and $P_n^r\{(L \otimes_k k', x \otimes 1)\}$ are isomorphic over K'. But because of remark 7.1 we have

$$P_n^r\{(L \otimes_k k', x \otimes 1)\}$$

$$= \mathrm{Tr}_{\left[L[X_j^{(i)}] \otimes_k k'\right]/\left[k[X_j^{(i)}] \otimes_k k'\right]} \left[\frac{1}{x \otimes 1} \left(\sum_{i,j} (e_j^{(i)} \otimes 1)(X_j^{(i)} \otimes 1)\right)^r\right]$$

$$\overset{7.1(2)}{=} \left(\mathrm{Tr}_{L[X_j^{(i)}]/k[X_j^{(i)}]} \left[\frac{1}{x} \left(\sum_{i,j} e_j^{(i)} X_j^{(i)}\right)^r\right]\right) \otimes 1$$

$$= P_n^r\{(L,x)\} \otimes 1,$$

and the corollary is proven. □

Corollary 7.8   *Again, let $r \geq 3$, and consider an $\mathcal{F}_K^{n,r}/\mathcal{F}_k^{n,r}$-form $P_n^r\{(L,x)\}$ of P_n^r. Then $P_n^r\{(L,x)\}$ and P_n^r become isomorphic over a field extension k' of k if and only if $L \otimes_k k'$ is isomorphic to k'^n over k' and $x \otimes 1$ is an r-th power in $L \otimes_k k'$.*

Proof.   Corollary 7.7 states that $P_n^r\{(L,x)\}$ and P_n^r become isomorphic over k' if and only if $(L \otimes_k k', x \otimes 1)$ and $(k^n \otimes_k k', 1 \otimes 1) = (k'^n, 1)$ define the same class in $H_{\mathrm{cont}}^1(G_{k'}, \mu_r \wr S_n)$. According to corollary 7.4, this is the case if and only if $L \otimes_k k'$ is isomorphic to k'^n over k' and $x \otimes 1$ is an r-th power in $L \otimes_k k'$. □

After having computed $E(\mathcal{F}_K^{n,r}/\mathcal{F}_k^{n,r}, P_n^r)$ respectively $H_{\mathrm{cont}}^1(G, A(P_n^r))$, we are now able to apply proposition 6.1 to determine the automorphism groups of twisted Fermat equations.

Corollary 7.9   *Let $r \geq 3$, let $b \in H_{cont}^1(G, \mu_r \wr S_n)$ be an arbitrary class, given by a pair (L,x) in the sense of corollary 7.4 with $L = \prod_{i=1}^n L_i$, and let $P_n^r\{b\}$ be the twisted Fermat equation defined in proposition 7.3. Then we have the following short exact sequence of groups:*

$$1 \to \prod_{i=1}^m (L_i \cap \mu_r) \to Aut_k\left(P_n^r\{b\}\right) \to \left\{a \in Aut_k(L)^{opp} \,\middle|\, \frac{ax}{x} \in L^{\times r}\right\} \to 1$$

$$(7.4)$$

In the special case $x = 1$, sequence (7.4) splits, and we get

$$Aut_k\left(P_n^r\{b\}\right) \cong \left(\prod_{i=1}^{m}(L_i \cap \mu_r)\right) \rtimes Aut_k(L)^{opp} \qquad (7.5)$$

Here the semidirect product is given by the following action:

$$Aut_k(L)^{opp} \times \prod_{i=1}^{m}(L_i \cap \mu_r) \longrightarrow \prod_{i=1}^{m}(L_i \cap \mu_r) \qquad (7.6)$$

$$(a \quad , \quad x) \qquad\qquad \mapsto \qquad a^{-1}(x)$$

Proof. Starting with the exact sequence

$$1 \longrightarrow \mu_r^n \longrightarrow \mu_r \wr S_n \longrightarrow S_n \longrightarrow 1,$$

twisting with the 1-cocycle b and applying corollary 4.5, we first get the following exact sequence of pointed sets:

$$H^0(G, (\mu_r^n)_b) \hookrightarrow H^0(G, (\mu_r \wr S_n)_b) \to H^0(G, (S_n)_b) \xrightarrow{\delta} H^1_{cont}(G, (\mu_r^n)_b).$$

Denote the image of b in $Z^1_{cont}(G, S_n)$ by c. Then $(S_n)_b = (S_n)_c$, and because of proposition 4.2(4) we also have $(\mu_r^n)_b = (\mu_r^n)_c$. In addition to that, we can identify $H^0(G, (\mu_r \wr S_n)_b)$ with $Aut_k(P_n^r\{b\})$ because of proposition 6.1. Corollary 7.2 and lemma 7.6 then imply that the following diagram of pointed sets commutes and has exact rows:

$$
\begin{array}{ccccccc}
H^0(G, (\mu_r^n)_c) & \hookrightarrow & H^0(G, (\mu_r \wr S_n)_b) & \twoheadrightarrow & H^0(G, (S_n)_c) & \xrightarrow{\delta} & H^1_{cont}(G, (\mu_r^n)_c) \\
\wr \uparrow \eta & & \wr \uparrow & & \wr \uparrow \alpha & & \wr \uparrow \gamma \\
\prod_{i=1}^{m}(L_i \cap \mu_r) & \hookrightarrow & Aut_k\left(P_n^r\{b\}\right) & \longrightarrow & Aut_k(L)^{opp} & \dashrightarrow_{\tilde{\delta}} & L^\times / L^{\times^r}
\end{array}
$$

We even know that all maps in this diagram except α, γ, δ and $\tilde{\delta}$ are *group homomorphisms*. In order to prove that (7.4) is an exact sequence of groups, we therefore only have to show that

$$\mathrm{Ker}\,(\tilde{\delta}) = \left\{a \in Aut_k(L)^{opp} \,\Big|\, \frac{ax}{x} \in L^{\times^r}\right\}.$$

This would obviously follow if we could show that $\tilde{\delta}$ is given by the formula

$$\tilde{\delta} : \mathrm{Aut}_k(L)^{\mathrm{opp}} \longrightarrow L^\times/L^{\times r}$$
$$a \mapsto \frac{ax}{x},$$
(7.7)

which can easily be checked — let $b = (b_s) = (a_s \cdot c_s)$ with $a_s \in \mu_r^n$ and $c_s \in \mathrm{Aut}(M)$:

$$\delta\alpha(a) \overset{\text{lemma 7.6}}{=} \delta[\varphi \mapsto \varphi \circ a]$$

$$\overset{\text{proposition 4.1}}{=} \left([\varphi \mapsto \varphi \circ a]^{-1} \cdot {}^{s'}[\varphi \mapsto \varphi \circ a] \right)_s$$

$$= \left([\varphi \mapsto \varphi \circ a^{-1}] \cdot a_s \cdot c_s \cdot [\varphi \mapsto \varphi \circ a] \cdot c_s^{-1} \cdot a_s^{-1} \right)_s$$

$$\overset{(4.13)}{=} \left([\varphi \mapsto \varphi \circ a^{-1}] \cdot a_s \cdot [\varphi \mapsto s \circ s^{-1} \circ \varphi \circ a] \cdot a_s^{-1} \right)_s$$

$$= \left({}^{[\varphi \mapsto \varphi \circ a^{-1}]}a_s \cdot \underbrace{[\varphi \mapsto s \circ s^{-1} \circ \varphi \circ a \circ a^{-1}]}_{=1_{\mathrm{Aut}(M)}} \cdot a_s^{-1} \right)_s$$

$$= \left({}^{[\varphi \mapsto \varphi \circ a^{-1}]}a_s \cdot a_s^{-1} \right)_s$$

$$\overset{\text{corollary 7.4}}{=} \left[\left(\frac{s\sqrt[r]{s^{-1}\varphi a x}}{\sqrt[r]{\varphi a x}} \cdot \frac{\sqrt[r]{\varphi x}}{s\sqrt[r]{s^{-1}\varphi x}} \right)_{\varphi \in M} \right]_s$$

$$= \left[\left(\frac{s\sqrt[r]{s^{-1}\varphi\left(\frac{ax}{x}\right)}}{\sqrt[r]{\varphi\left(\frac{ax}{x}\right)}} \right)_{\varphi \in M} \right]_s$$

$$\overset{\text{corollary 7.1}}{=} \gamma\left(\frac{ax}{x} \right).$$

Formula (7.7) indeed follows, because $\delta\alpha = \gamma\tilde{\delta}$ and because γ is an isomorphism. So the exactness of sequence (7.4) is proven.

Now consider the special case $x = 1$. Then b is contained in the image of the map $Z^1_{\mathrm{cont}}(G, S_n) \hookrightarrow Z^1_{\mathrm{cont}}(G, \mu_r \wr S_n)$, which is induced by the canonical section $S_n \hookrightarrow \mu_r \wr S_n$, and (7.5) follows from proposition 4.2(3).

Finally, (7.6) follows from the following diagram of actions which obviously commutes:

$$\left(\sigma \qquad , \qquad (\zeta_\varphi)_{\varphi \in M}\right) \quad \longmapsto \quad (\zeta_{\sigma^{-1}\varphi})_{\varphi \in M}$$

$$
\begin{array}{ccccccc}
[\varphi \mapsto \varphi a] & \mathrm{Aut}(M) & \times & \mu_r^n & \longrightarrow & \mu_r^n & (\varphi x)_{\varphi \in M} \\[2mm]
& \uparrow & & \uparrow & & \uparrow & \\[2mm]
& H^0(G,(S_n)_c) & \times & H^0(G,(\mu_r^n)_b) & \longrightarrow & H^0(G,(\mu_r^n)_b) & \\[2mm]
& \alpha \mid \wr & & \eta \mid \wr & & \eta \mid \wr & \\[2mm]
a & \mathrm{Aut}_k(L)^{\mathrm{opp}} & \times & \displaystyle\prod_{i=1}^m (L_i \cap \mu_r) & \longrightarrow & \displaystyle\prod_{i=1}^m (L_i \cap \mu_r) & x
\end{array}
$$

$$\left(a \qquad , \qquad x\right) \quad \longmapsto \quad a^{-1}(x)$$

$\square$

Remark 7.2 *We deduced sequences (7.4) and (7.5) from our general formalism. As Sladek and Wesołowski showed in [SW98, theorem 1.3.] respectively [Wes99, theorem 3.3.], they can also be proven elementary by direct calculation.*

Example 7.3 Consider the special case $n = 6$, $r = 3$, $k = \mathbb{F}_7$ and $K = \overline{\mathbb{F}}_7$ from example 7.1.

As in that example, we write $\mathbb{F}_{49}$ as $\mathbb{F}_7(\alpha)$ with $\alpha^2 + 5\alpha + 5 = 0$ and $\mathbb{F}_{2401}$ as $\mathbb{F}_7(\beta)$ with $\beta^4 + 5\beta^3 + 4\beta^2 + \beta + 5 = 0$, and we consider the class $b \in H^1_{\mathrm{cont}}(G, \mu_3 \wr S_6)$ given by the pair $\left(\mathbb{F}_{2401} \times \mathbb{F}_{49}, (\frac{1}{\beta}, \frac{1}{\alpha^2})\right)$.

In order to determine $P_6^3\{b\}$, we first have to do some calculations in the fields $\mathbb{F}_{49}$ and $\mathbb{F}_{2401}$ to compute the necessary traces:

$$\alpha^7 = 6\alpha + 2 \implies \mathrm{Tr}_{\mathbb{F}_{49}/\mathbb{F}_7}(\alpha) = \alpha + \alpha^7 = 2,$$

$$\left.\begin{array}{l} \beta^7 \ \ = 3\beta^3 + 5\beta^2 + 2\beta + 3 \\ \beta^{49} \ = 2\beta^3 + \beta + 6 \\ \beta^{343} = 2\beta^3 + 2\beta^2 + 3\beta \end{array}\right\} \implies \mathrm{Tr}_{\mathbb{F}_{2401}/\mathbb{F}_7}(\beta) = 2,$$

$$\left.\begin{array}{l} (\beta^2)^7 \ \ = 3\beta^3 + \beta^2 + 4\beta \\ (\beta^2)^{49} \ = 3\beta^3 + 6\beta^2 + 3\beta + 2 \\ (\beta^2)^{343} = \beta^3 + 6\beta^2 + 1 \end{array}\right\} \implies \mathrm{Tr}_{\mathbb{F}_{2401}/\mathbb{F}_7}(\beta^2) = 3,$$

$$\left.\begin{array}{l} (\beta^3)^7 \ \ = 3\beta^3 + 2\beta^2 + 4\beta + 6 \\ (\beta^3)^{49} \ = 6\beta^3 + 1 \\ (\beta^3)^{343} = 4\beta^3 + 5\beta^2 + 3\beta + 2 \end{array}\right\} \implies \mathrm{Tr}_{\mathbb{F}_{2401}/\mathbb{F}_7}(\beta^3) = 2.$$

For the following computation of $P_6^3\{b\}$, we want to use the variable names a, b, c, d, x and y instead of $X_1^{(1)},\dots,X_4^{(1)}$, $X_1^{(2)}$ and $X_2^{(2)}$, because it is shorter and simpler that way. We use $\{1,\alpha\}$ as an $\mathbb{F}_7$-basis of $\mathbb{F}_{49}$ and $\{1,\beta,\beta^2,\beta^3\}$ as an $\mathbb{F}_7$-basis of $\mathbb{F}_{2401}$. With this we get:

$$\begin{aligned} P_6^3\{b\} &= \mathrm{Tr}_{\mathbb{F}_{2401}/\mathbb{F}_7}\left[\beta \cdot \left(a + \beta b + \beta^2 c + \beta^3 d\right)^3\right] \\ &\qquad\qquad\qquad + \mathrm{Tr}_{\mathbb{F}_{49}/\mathbb{F}_7}\left[\alpha^2 \cdot (x + \alpha y)^3\right] \\ &= 2a^3 + 6a^2 c + a^2 d + 6ab^2 + 2abc + 4abd + 2ac^2 + 3acd + 6ad^2 \\ &\quad + 5b^3 + 2b^2 c + 5b^2 d + 5bc^2 + 5bcd + 6bd^2 \\ &\quad + 2c^3 + 6c^2 d + cd^2 + d^3 \\ &\quad + y^3 + 4x^2 y + 5y^3. \end{aligned}$$

Example 7.4   Now we want to continue example 7.2 and compute the forms corresponding to the cohomology classes listed there:

(1) $k = \mathbb{R}$, $K = \mathbb{C}$, $n = 3$ and $r = 4$.

The only separable field extension of $\mathbb{R}$ is $\mathbb{C}$, and we choose $\{1,i\}$ as $\mathbb{R}$-basis of $\mathbb{C}$. In terms of this basis, the trace of $\mathbb{C}$ over $\mathbb{R}$ looks as follows:

$$\mathrm{Tr}_{\mathbb{C}/\mathbb{R}}(a + ib) = (a + ib) + (a - ib) = 2a,$$

and we get (for simplicity we use variables $x, y, \dots$ instead of $X_1^{(1)}, X_2^{(1)}, \dots$):

b	$P_3^4\{b\}$
$\left(\mathbb{R}^3,(1,1,1)\right)$	$x^4 + y^4 + z^4$
$\left(\mathbb{R}^3,(1,1,-1)\right)$	$x^4 + y^4 - z^4$
$\left(\mathbb{R}^3,(1,-1,-1)\right)$	$x^4 - y^4 - z^4$
$\left(\mathbb{R}^3,(-1,-1,-1)\right)$	$-x^4 - y^4 - z^4$
$\left(\mathbb{C}\times\mathbb{R},(1,1)\right)$	$(2x^4 - 12x^2y^2 + 2y^4) + z^4$
$\left(\mathbb{C}\times\mathbb{R},(1,-1)\right)$	$(2x^4 - 12x^2y^2 + 2y^4) - z^4.$

(2) $k = \mathbb{F}_5$, $K = \overline{\mathbb{F}}_5$, $n = 4$ and $r = 3$.

We know from example 7.2 that $\mathbb{F}_{25} = \mathbb{F}_5(\alpha)$ with $\alpha^2 + 2\alpha + 3 = 0$, and we want to take $\{1,\alpha\}$ as $\mathbb{F}_5$-basis of $\mathbb{F}_{25}$. Then for the trace we get:

$$\mathrm{Tr}_{\mathbb{F}_{25}/\mathbb{F}_5}(a + \alpha b) = 2a + 3b.$$

We get the field $\mathbb{F}_{125}$ by adjoining γ with $\gamma^3 + 4\gamma + 3 = 0$ to $\mathbb{F}_5$; then we can take $\{1,\gamma,\gamma^2\}$ as $\mathbb{F}_5$-basis of $\mathbb{F}_{125}$, and get for the trace:

$$\mathrm{Tr}_{\mathbb{F}_{125}/\mathbb{F}_5}(a + \gamma b + \gamma^2 c) = 3a + 4b + 2c.$$

As mentioned earlier, we have $\mathbb{F}_{625} = \mathbb{F}_5(\beta)$ with $\beta^4 + \beta^2 + 3\beta + 3 = 0$. If we choose $\{1,\beta,\beta^2,\beta^3\}$ as $\mathbb{F}_5$-basis of $\mathbb{F}_{625}$, we get the following formula for the trace:

$$\mathrm{Tr}_{\mathbb{F}_{625}/\mathbb{F}_5}(a + \beta b + \beta^2 c + \beta^3 d) = 4a + 3c + d.$$

Using these formulas, we can now give a complete list of the forms of P_4^3 over $\mathbb{F}_5$:

b	$P_4^3\{b\}$
$\left(\mathbb{F}_5^4, (1,1,1,1)\right)$	$x^3 + y^3 + z^3 + u^3$
$\left(\mathbb{F}_{25} \times \mathbb{F}_5^2, (1,1,1)\right)$	$(2x^3 + 4x^2y + 4xy^2) + z^3 + u^3$
$\left(\mathbb{F}_{25} \times \mathbb{F}_5^2, (\alpha,1,1)\right)$	$(3x^3 + 4x^2y + y^3) + z^3 + u^3$
$\left(\mathbb{F}_{25}^2, (1,1)\right)$	$(2x^3 + 4x^2y + 4xy^2) + (2z^3 + 4z^2u + 4zu^2)$
$\left(\mathbb{F}_{25}^2, (1,\alpha)\right)$	$(2x^3 + 4x^2y + 4xy^2) + (3z^3 + 4z^2u + u^3)$
$\left(\mathbb{F}_{25}^2, (\alpha,\alpha)\right)$	$(3x^3 + 4x^2y + y^3) + (3z^3 + 4z^2u + u^3)$
$\left(\mathbb{F}_{125} \times \mathbb{F}_5, (1,1)\right)$	$(3x^3 + xy^2 + 2x^2y + x^2z + 2yz^2) + u^3$
$\left(\mathbb{F}_{625}, 1\right)$	$x^3 + 4xy^2 + y^3 + 4x^2z + 3x^2u + xyz + 4xu^2$ $\qquad +4yzu + yu^2 + 3z^3 + z^2u + zu^2 + 4u^3$
$\left(\mathbb{F}_{625}, \beta\right)$	$4x^2y + 3xy^2 + 3x^2z + 4y^2u + 4xzu + xu^2$ $\qquad +4yz^2 + yzu + zu^2 + 2z^3 + z^2u + 2zu^2 + 3u^3$

Chapter 8

Binary cubic equations

In this chapter, we want to apply the results from the last chapter to the special case $n = 2$ and $r = 3$, i.e. to the case of *binary cubic equations*. This case will prove to be of particular interest for two reasons: First, we will see that all non-singular binary cubic equations are forms of the Fermat equation $P := P_2^3 = X^3 + Y^3$, so that our methods allow us to study *all* these equations. Second, in this case we can not only explicitly compute ϑ^{-1}, but ϑ itself as well, which means that we will be able to give an explicit formula that for a given non-singular binary cubic equation Q computes the pair $(L, x) = \vartheta[Q]$.

Let k be a field whose characteristic is neither two nor three, let K be its separable closure, and let G be the absolute Galois group $\mathrm{Gal}(K/k)$.

Definition 8.1 A *binary cubic equation* (over k) is an object of $\mathcal{F}_k^{2,3}$, i.e. a homogenous polynomial of degree three in two variables. The *discriminant* of a binary cubic equation $Q(X, Y) = aX^3 + bX^2Y + cXY^2 + dY^3$ is defined as

$$\Delta(Q) := -27a^2d^2 + 18abcd + b^2c^2 - 4b^3d - 4ac^3, \qquad (8.1)$$

and Q is called *non-singular* if $\Delta(Q) \neq 0$. Let $\mathcal{F} \subset \mathcal{F}_k^{2,3}$ denote the full subcategory of non-singular equations. In particular, the Fermat equation $P_2^3 = X^3 + Y^3$ is non-singular because

$$\Delta(P_2^3) = -27 \cdot 1 + 18 \cdot 0 + 0 - 4 \cdot 0 - 4 \cdot 0 = -27 \neq {}^1 0.$$

Remark 8.1 *In [GKZ94], Gelfand and Kapranov associate a discrimi-*

---

[1] Remember that we assume the characteristic of k to be neither two nor three.

nant $\Delta(f)$ to every polynomial f of arbitrary degree and in arbitrary many variables. The above defined discriminant of a binary cubic equation Q is $\Delta(Q(1,X))$ in the sense of [GKZ94].

In particular, the general theory implies that Q is non-singular if and only if the associated hypersurface[2] $X(Q) := \tilde{F}_k F_k Q$ in $\mathbb{P}^1_k$ is smooth.

Lemma 8.1   Let $Q(X,Y)$ be an arbitrary binary cubic equation over k.

(1) For $A = \left(\begin{smallmatrix} r & s \\ t & u \end{smallmatrix}\right) \in GL(2,k)$ we have

$$\Delta\underbrace{\left(Q(rX+sY, tX+uY)\right)}_{=:Q_A} = \left[\det(A)\right]^6 \cdot \Delta(Q);$$

in particular, the discriminant induces a well defined map

$$\left\{ \text{isomorphism classes in } \mathcal{F} \right\} \xrightarrow{\Delta} k^\times / k^{\times 6}.$$

(2) If in $K[X,Y]$ we have $Q(X,Y) = a(X-\alpha Y)(X-\beta Y)(X-\gamma Y)$, then

$$\Delta(Q) = a^4 (\alpha-\beta)^2 (\alpha-\gamma)^2 (\beta-\gamma)^2.$$

(3) Q is a $\mathcal{F}_K^{2,3}/\mathcal{F}_k^{2,3}$-form of the Fermat equation P if and only if Q is non-singular.

Proof.   Parts (1) and (2) either follow from the general properties of discriminants as described in [GKZ94] or can be checked directly by a simple computation.

Then (1) implies that any $\mathcal{F}_K^{2,3}/\mathcal{F}_k^{2,3}$-form of P_2^3 has to be non-singular. To prove (3), let us therefore assume that $\Delta(Q) \neq 0$.

The hypersurface $X(Q) \subset \mathbb{P}^1_k = \mathrm{Proj}\,(k[X,Y])$ is zero-dimensional and consists of at most three K-rational points. It is a well known fact (analogous to the classical case of Moebius transformations of the Riemann sphere $\hat{\mathbb{C}}$) that to any two given triples of distinct points in $\mathbb{P}^1_k(K) = \mathbb{P}^1_K(K)$, there is a matrix in $GL(2,K) = \mathrm{Aut}(\mathbb{P}^1_K)$ which sends the points of the first triple to those in the second triple. Therefore we can find a matrix A that maps all points of $X(Q)$ to points that are distinct from $[1:0]$.

What does that say about the equation $Q_A = aX^3 + bX^2Y + cXY^2 + dY^3$?

---

[2]For the definitions of the functors F_k and $\tilde{F}_k$, compare example 3.4.

— If $[1:0]$ is not a solution of Q_A, then $a \cdot 1 + b \cdot 0 + c \cdot 0 + d \cdot 0 \neq 0$, i.e. $a \neq 0$. We then get (for suitable elements α, β and γ of K):

$$
\begin{aligned}
Q_A &= aX^3 + bX^2Y + cXY^2 + dY^3 \\
&\stackrel{a \neq 0}{=} aY^3 \cdot \left(\left[\tfrac{X}{Y}\right]^3 + \tfrac{b}{a} \cdot \left[\tfrac{X}{Y}\right]^2 + \tfrac{c}{a} \cdot \left[\tfrac{X}{Y}\right] + \tfrac{d}{a} \right) \\
&= aY^3 \cdot \left(\left[\tfrac{X}{Y}\right] - \alpha \right) \left(\left[\tfrac{X}{Y}\right] - \beta \right) \left(\left[\tfrac{X}{Y}\right] - \gamma \right) \\
&= a(X - \alpha Y)(X - \beta Y)(X - \gamma Y).
\end{aligned}
$$

Because Q and Q_A are non-singular, (2) implies that α, β and γ are three distinct elements of K. So we can find a $B \in \mathrm{GL}(2, K)$ which maps α, β and γ to (-1), $(-\zeta)$ and $(-\zeta^2)$ for a primitive third root of unity $\zeta \in K$. Put $C := \mathrm{diag}\left(\tfrac{1}{\sqrt[3]{a}}, \tfrac{1}{\sqrt[3]{a}} \right) \in \mathrm{GL}(2, K)$, then we get:

$$
Q_{CBA} = a \cdot \underbrace{\left(\frac{1}{\sqrt[3]{a}} \right)^3}_{=1} \cdot (X + Y)(X + \zeta Y)(X + \zeta^2 Y)
$$

$$
= X^3 + \underbrace{(\zeta^2 + \zeta + 1)}_{=0} X^2Y + \underbrace{(\zeta^3 + \zeta^2 + \zeta)}_{=0} XY^2 + \zeta^3 Y^3 = P,
$$

i.e. CBA defines an isomorphism from P to Q in $\mathcal{F}_K^{2,3}$, so that Q is indeed a $\mathcal{F}_K^{2,3}/\mathcal{F}_k^{2,3}$-form of P. □

Definition 8.2 For $\delta \in k^\times / k^{\times 2}$ define

$$
L_\delta := \begin{cases} k \times k & \text{if } \delta \in k^{\times 2}, \\ k(\sqrt{\delta}) & \text{otherwise.} \end{cases}
$$

Then $\delta \mapsto L_\delta$ obviously defines a bijection between $k^\times / k^{\times 2}$ and the set of isomorphism classes of fcs-algebras of degree two over k.

Finally, put $A_\delta := \mathrm{Aut}_k(L_\delta)$ and denote the generator of this group of order two by τ, i.e. we have $\tau(x, y) = (y, x)$ for $\delta \in k^{\times 2}$ and $\tau(x + y\sqrt{\delta}) = x - y\sqrt{\delta}$ otherwise.

Corollary 8.1   *We have a canonical bijection*

$$
\coprod_{\delta \in k^\times / k^{\times 2}} \left[A_\delta \backslash \left(L_\delta^\times / L_\delta^{\times 3} \right) \right] \longrightarrow \left\{ \text{isomorphism classes in } \mathcal{F} \right\}
$$
$$
(\delta, x) \qquad\qquad \mapsto \qquad\qquad P\left\{ (L_\delta, x) \right\}.
$$

Proof.   From lemma 8.1(3) we know that the set of isomorphism classes in $\mathcal{F}$ is the same as the set of $\mathcal{F}_K^{2,3} / \mathcal{F}_k^{2,3}$-forms of P, and corollary 7.6 states that ϑ induces a bijection from this set to the set $H^1_{\text{cont}}(G, \mu_3 \wr S_2)$. Now, according to corollary 7.4, we have

$$
H^1_{\text{cont}}(G, \mu_3 \wr S_2) \cong \coprod_{\substack{\text{isomorphism classes} \\ \text{of fcs-algebras } L \\ \text{of degree 2 over } k}} \mathrm{Aut}_k(L) \backslash \left[L^\times / L^{\times 3} \right],
$$

and the corollary follows from the remark in definition 8.2.   □

Corollary 8.2   *Consider the special case where $k = \mathbb{F}_q$ is a finite field, and let Q be an arbitrary non-singular binary cubic equation over k.*

Put $k' := \mathbb{F}_{q^2}$, choose a generator α of the multiplicative group $k'^\times$, and put $\beta := \alpha^{\frac{q+1}{2}}$ and $\delta := \beta^2$. Then:

(1)   If k contains the third roots of unity, or equivalently if $q \equiv 1 \pmod 3$, then Q is isomorphic to exactly one of the following nine twisted Fermat equations:

$$
\begin{aligned}
P_2^3\{(L_1, (1,1))\} &= X^3 + Y^3, \\
P_2^3\{(L_1, (1,\delta))\} &= X^3 + \delta Y^3, \\
P_2^3\{(L_1, (1,\delta^2))\} &= X^3 + \delta^2 Y^3, \\
P_2^3\{(L_1, (\delta,\delta))\} &= \delta X^3 + \delta Y^3, \\
P_2^3\{(L_1, (\delta,\delta^2))\} &= \delta X^3 + \delta^2 Y^3, \\
P_2^3\{(L_1, (\delta^2,\delta^2))\} &= \delta^2 X^3 + \delta^2 Y^3, \\[4pt]
P_2^3\{(L_\delta, 1)\} &= 2X^3 + 6\delta XY^2, \\
P_2^3\{(L_\delta, \alpha)\} &= \mathrm{Tr}_{k'/k}(\alpha)X^3 + 3\,\mathrm{Tr}_{k'/k}(\alpha\beta)X^2 Y \\
&\quad + 3\delta\,\mathrm{Tr}_{k'/k}(\alpha)XY^2 + \delta\,\mathrm{Tr}_{k'/k}(\alpha\beta)Y^3, \\
P_2^3\{(L_\delta, \alpha^2)\} &= \mathrm{Tr}_{k'/k}(\alpha^2)X^3 + 3\,\mathrm{Tr}_{k'/k}(\alpha^2\beta)X^2 Y \\
&\quad + 3\delta\,\mathrm{Tr}_{k'/k}(\alpha^2)XY^2 + \delta\,\mathrm{Tr}_{k'/k}(\alpha^2\beta)Y^3.
\end{aligned}
$$

*(2) If k does not contain the third roots of unity, or equivalently if $q \equiv 2$
(mod 3), then Q is isomorphic to exactly one of the following three
twisted Fermat equations:*

$$
\begin{aligned}
P_2^3\{(L_1,(1,1))\} &= X^3 + Y^3, \\[4pt]
P_2^3\{(L_\delta,1)\} &= 2X^3 + 6\delta XY^2 \\
P_2^3\{(L_\delta,\alpha)\} &= Tr_{k'/k}(\alpha)X^3 + 3\,Tr_{k'/k}(\alpha\beta)X^2Y \\
&\quad + 3\delta\, Tr_{k'/k}(\alpha)XY^2 + \delta\, Tr_{k'/k}(\alpha\beta)Y^3.
\end{aligned}
$$

Proof.   The index of $k^\times$ in $k'^\times$ is $\frac{q^2-1}{q-1} = q+1$, so $\delta = \beta^2 = \alpha^{q+1}$ generates
$k^\times$, because α generates $k'^\times$. It follows that $[\delta]$ generates $k^\times/k^{\times 3}$ and that
δ can not be a square in k, so that $k' = \mathbb{F}_q(\sqrt{\delta}) = L_\delta$. Put $H := L_\delta^\times/L_\delta^{\times 3}$
— it is obvious that $[\alpha]$ is a generator of H.

First consider the case $q \equiv 1$ (mod 3), i.e. $q = 3t + 1$ for a suitable $t \in \mathbb{N}_+$.
Then the third roots of unity are of course also contained in L_δ, so that we
have an exact sequence of abelian groups

$$1 \to \mu_3 \to L_\delta^\times \xrightarrow{3} L_\delta^\times \to H \to 1,$$

which implies that H contains three elements, namely those represented by
1, α and α^2. We have

$$\tau[\alpha] = \alpha^q = \alpha^{3t+1} = \alpha \cdot (\alpha^t)^3,$$

which means that A_δ acts trivially on H, so that there are three orbits in
H and thus three elements in $A_\delta \backslash H$, of course again represented by 1, α
and α^2.

Now look at $A_1 \backslash [L_1^\times / L_1^{\times 3}]$. Because we are in the case $\mu_3 \leq k^\times$, we can
conclude as before that $k^\times/k^{\times 3}$ consists of three elements, and since we have
seen above that $[\delta]$ generates $k^\times/k^{\times 3}$, we know that these three elements
are those represented by 1, δ and δ^2. Furthermore we know from lemma 7.7
that $A_1 = S_2$, generated by the automorphism $(x,y) \mapsto (y,x)$ of $L_1 = k \times k$.
From these observations it is clear that

$$A_1 \backslash \left[L_1^\times / L_1^{\times 3} \right] = \left\{ (1,1), (1,\delta), (1,\delta^2), (\delta,\delta), (\delta,\delta^2), (\delta^2,\delta^2) \right\}.$$

So according to corollary 8.1, in this case there are the following nine isomorphism classes in $\mathcal{F}$:

$P_2^3\{(L_\delta,1)\}$, $P_2^3\{(L_\delta,\alpha)\}$, $P_2^3\{(L_\delta,\alpha^2)\}$, $P_2^3\{(L_1,(1,1))\}$, $P_2^3\{(L_1,(1,\delta))\}$, $P_2^3\{(L_1,(1,\delta^2))\}$, $P_2^3\{(L_1,(\delta,\delta))\}$, $P_2^3\{(L_1,(\delta,\delta^2))\}$ and $P_2^3\{(L_1,(\delta^2,\delta^2))\}$.

The explicit formulas for these nine equations as stated in (1) can easily be computed from formula (7.3) if we choose $\{1,\beta\}$ as k-basis of L_δ and $\{(1,0),(0,1)\}$ as k-basis of L_1.

Consider now the case $q \equiv 2 \pmod 3$ where $q = 3t + 2$ for a suitable $t \in \mathbb{N}_+$. Then $q^2 \equiv 1 \pmod 3$, so that $L_\delta^\times$ again contains μ_3 and again we have $\#H = 3$, but

$$\tau[\alpha] = \alpha^q = \alpha^{3t+2} = \alpha^2 \cdot (\alpha^t)^3,$$

which means that A_δ fixes the element $[1]$ of H and swaps the elements $[\alpha]$ and $[\alpha^2]$. Consequently there are only two orbits in H under the action of A_δ, and $A_\delta\backslash[L_\delta^\times/L_\delta^{\times3}]$ only contains two elements this time, represented by 1 and α. Furthermore, in this case $k^\times \xrightarrow{3} k^\times$ is bijective, so that $k^\times/k^{\times3}$ and $L_1^\times/L_1^{\times3}$ are singleton sets. So now we only get three isomorphism classes in $\mathcal{F}$, namely $P_2^3\{(L_\delta,1)\}$, $P_2^3\{(L_\delta,\alpha)\}$ and $P_2^3\{(L_1,(1,1))\}$. We find the explicit formulas stated in (2) by using the same bases for L_δ and L_1 as in the first case. □

Corollary 8.1 states that isomorphism classes of non-singular binary cubic equations are classified by pairs (δ,x) with $\delta \in k^\times/k^{\times2}$ and $x \in A_\delta\backslash\left[L_\delta/L_\delta^{\times3}\right]$, and for a given such pair (δ,x) we can use formula (7.3) to compute the corresponding equation.

Of course it would be nice to also have a formula that computes δ and x to a given equation. As already mentioned in the introduction to this chapter, it is indeed possible to give such a formula, which we are going to do in theorem 8.1. But before, we need the following lemma:

Lemma 8.2　*Let $Q(X,Y)$ be a non-singular cubic equation over k. Then there exists an $A \in GL(2,k)$ such that*

$$Q_A(X,Y) = a \cdot \left(X^3 + bX^2Y + cXY^2 + dY^3\right) \tag{8.2}$$

with $a \in k^\times$ and $b,c,d \in k$, and we can even choose A to be of the form $A = \left(\begin{smallmatrix} 1 & 0 \\ \rho & 1 \end{smallmatrix}\right)$ with $\rho \in k$.

If M denotes the splitting field of

$$f(X) := \frac{1}{a} \cdot Q_A(X,1) = X^3 + bX^2 + cX + d \in k[X],$$

then M is independent of the choice of A, and the discriminant of f does not vanish.

Proof. Let $Q(X,Y) = a'X^3 + b'X^2Y + c'XY^2 + d'Y^3$. If $a' \neq 0$, then $Q(X,Y)$ is already in the right form, and we can choose $\rho := 0$.

Let us therefore assume that $a' = 0$. Then because Q is non-singular, b', c' and d' can not all be zero, i.e. we have $P(X) := d'X^3 + c'X^2 + b'X \neq 0 \in k[X]$. Since k contains more than three elements (remember, k is not of characteristic two or three and can therefore be neither $\mathbb{F}_2$ nor $\mathbb{F}_3$), we find a $\rho \in k$ with $P(\rho) \neq 0$ and get

$$Q_{\left(\begin{smallmatrix} 1 & 0 \\ \rho & 1 \end{smallmatrix}\right)}(X,Y)$$
$$= \underbrace{(d'\rho^3 + c'\rho^2 + b'\rho)}_{P(\rho)\neq 0} X^3 + (b' + 2c'\rho + 3d'\rho^2)X^2Y$$
$$+ (c' + 3d'\rho)XY^2 + d'Y^3,$$

which is in the right form.

Let α_1, α_2 and α_3 denote the roots of f in K. Then

$$Q_A(X,Y) = a \cdot \prod_{i=1}^{3} (X - \alpha_i Y),$$

and because Q_A is non-singular, lemma 8.1(2) implies that the α_i are distinct, i.e. the discriminant of f does not vanish.

If $A^* \in \mathrm{GL}(2,k)$ is another matrix satisfying

$$Q_{A^*}(X,Y) = a^* \cdot \left(X^3 + b^*X^2Y + c^*XY^2 + d^*Y^3\right)$$

with $a^* \in k^\times$ and $b^*, c^*, d^* \in k$, we can put $f^* := \frac{1}{a^*} \cdot Q(X,1)$ and $B :=$

$A^*A^{-1} = \begin{pmatrix} r & s \\ t & u \end{pmatrix}$ and get:

$$
\begin{aligned}
f^*(X) &= \frac{1}{a^*} \cdot (Q_A)_B(X,1) \\
&= \frac{a}{a^*} \cdot \prod_{i=1}^{3}\Big((rX+s) - \alpha_i(tX+u)\Big) \\
&= \frac{a}{a^*} \cdot \prod_{i=1}^{3}\Big((r - \alpha_i t)X - (\alpha_i u - s)\Big) \\
&\overset{a^* \neq 0}{=} \underbrace{\left(\frac{a}{a^*} \prod_{i=1}^{3}(r - \alpha_i t) \right)}_{=1} \cdot \prod_{i=1}^{3}\Big(X - \frac{\alpha_i u - s}{r - \alpha_i t}\Big).
\end{aligned}
$$

This shows that the roots of f^* are rational expressions in the α_i with coefficients in k, so the splitting field of f^* is contained in M. By symmetry, the two fields then have to be equal, which proves that M really does not depend on the choice of A.   □

Theorem 8.1   *Let $Q(X,Y) = aX^3 + bX^2Y + cXY^2 + dY^3$ be a non-singular equation over k with $a \neq 0$,[3] and put*

$$
\boxed{
\begin{aligned}
\delta(Q) := \delta &:= -\frac{\Delta(Q)}{27} && \in k^\times, \\
e(Q) := e &:= \frac{a}{2} - \frac{27a^2 d + 2b^3 - 9abc}{2\Delta(Q)}\sqrt{\delta} && \in k(\sqrt{\delta})^\times,
\end{aligned}
}
$$

where in the formula for e the squareroot of δ has to be chosen such that $e \neq 0$.[4]

Then

$$
\boxed{
Q \cong
\begin{cases}
P_2^3\left\{ \left(L_\delta, \left(e, \dfrac{\sqrt{\delta}}{e}\right) \right) \right\} & \text{if } \delta \in k^{\times^2}, \\[2ex]
P_2^3\left\{ (L_\delta, e) \right\} & \text{otherwise.}
\end{cases}
}
$$

---

[3] Note that the assumption "$a \neq 0$" is no restriction, because if $a = 0$, the we can find an isomorphic equation with $a \neq 0$ according to lemma 8.2.

[4] This is possible, because a is not zero. Therefore, if we got $e = 0$ for one choice of $\sqrt{\delta}$, we must get $e = \frac{a}{2} + \frac{a}{2} = a \neq 0$ for the other

Remark 8.2 *In [HM00], Hoffman and Morales study binary cubic equations over much more general ground rings R. Our corollary 8.1 is only a special case of their result, and also our formula $\delta = -\frac{\Delta(Q)}{27}$ is deduced there, it even goes back to classical results of Eisenstein.*

In contrast to that, the element e in theorem 8.1 is an interesting new invariant which in the case $R = k$ allows us to explicitly compute the canonical isomorphism ϑ from corollary 7.6.

For the proof of the theorem, we will need two technical lemmas. So let us start with a few remarks on classical abelian Galois cohomology which we will need to prove the first lemma:

- If R is a field with absolute Galois group G_R, then we write $H^i(R, A)$ instead of $H^i_{\text{cont}}(G_R, A)$ for all $i \geq 0$ and all abelian G_R-modules A.
- If R/S is a Galois extension of fields, then we write $H^i(R/S, A)$ instead of $H^i_{\text{cont}}(\text{Gal}(R/S), A)$ for all $i \geq 0$ and all abelian G_R-modules A.
- If $R/S/T$ is a tower of fields with R/T and S/T Galois, then the following inflation-restriction-sequence[5] is an exact sequence of abelian groups for all $\text{Gal}(R/T)$-modules A:

$$0 \to H^1(S/T, A^{\text{Gal}(R/S)}) \xrightarrow{\text{inf}} H^1(R/T, A) \xrightarrow{\text{res}} H^1(R/S, A).$$

  In particular, if R is the separable closure of S, this sequence looks as follows:

$$0 \to H^1(S/T, A^{G_S}) \xrightarrow{\text{inf}} H^1(T, A) \xrightarrow{\text{res}} H^1(S, A).$$

- If R/S is a finite Galois extension of fields of order β, if A is a finite abelian $\text{Gal}(R/S)$-module of order α, and if α and β are relatively prime, then $H^1(R/S, A) = 0$.
- If R is a field, if $r \in \mathbb{N}_+$ is a natural number, and if μ_r denotes the group of r-th roots of unity in a separable closure of R, then we have the canonical Kummer-isomorphism

$$R^\times / R^{\times^r} \xrightarrow{\sim} H^1(R, \mu_r), \quad \bar{\lambda} \mapsto \left(\frac{s\left[\sqrt[r]{\lambda}\right]}{\sqrt[r]{\lambda}} \right)_s. \tag{8.3}$$

---

[5]Compare proposition 4.4.

If R/S is a field extension, then the Kummer-isomorphism is compatible with restriction in the sense that the following diagram commutes:

$$
\begin{array}{ccc}
S^{\times}/S^{\times r} & \xrightarrow{\ [x]\mapsto[x]\ } & R^{\times}/R^{\times r} \\
\wr \downarrow & & \downarrow \wr \\
H^1(S,\mu_r) & \xrightarrow[\text{res}]{} & H^1(R,\mu_r)
\end{array}
$$

- If R/S is a Galois extension of fields, and if A is an abelian group endowed with the *trivial* $\mathrm{Gal}(R/S)$-action, then

$$
H^1(R/S, A) = \mathrm{Hom}_{\mathrm{Groups}}\big(\mathrm{Gal}(R/S), A\big).
$$

Lemma 8.3 *Let $f \in k[X]$ be a (not necessarily irreducible) monic polynomial of degree three over k whose discriminant Δ is not zero, put $\delta := -\frac{\Delta}{27}$ and $L := k(\sqrt{\delta})$, let M be the splitting field of f, and let ζ be a primitive third root of unity in K. Then:*

(1) There is an $\varepsilon \in L^{\times}$ with the property that ε is the third power of a $u \in M' := M(\zeta)$ and that $M' = L(\zeta, u)$.

(2) If $\varepsilon' \in L^{\times}$ is another element with $M' = L(\zeta, u')$ for a suitable $u' \in M'$ satisfying $(u')^3 = \varepsilon'$, then there is an $\eta \in L^{\times}$ such that $\varepsilon' = \varepsilon\eta^3$ or $\frac{1}{\varepsilon'} = \varepsilon\eta^3$. If δ is not a square in $k^{\times}$ (so that $[L : k] = 2$) and if τ denotes the non-trivial element of $\mathrm{Gal}(L/k)$, then there is an $\eta \in L^{\times}$ such that $\varepsilon' = \varepsilon\eta^3$ or $\tau(\varepsilon') = \varepsilon\eta^3$.

Proof.

(1) Consider an arbitrary $f = X^3 + bX^2 + cX + d \in k[X]$. For every $a \in k$, the polynomial $f(X - a)$ has the same discriminant and the same splitting field as f. In particular, if we choose $a = \frac{b}{3}$, we get

$$
f\left(X - \frac{b}{3}\right) = X^3 + \left(c - \frac{1}{3}b^2\right)X - \left(\frac{2}{27}b^3 - \frac{1}{3}bc + d\right),
$$

so that without loss of generality, we can assume $b = 0$. Then, putting $p := \frac{c}{3}$ and $q := \frac{d}{2}$, we get $f = X^3 + 3pX + 2q$.

Then $\Delta = -4 \cdot 27 \cdot (p^3 + q^2)$ and $\delta = 4(p^3 + q^2)$, and we put

$$\varepsilon := -q - \sqrt{p^3 + q^2} = -q - \frac{1}{2}\sqrt{\delta}, \qquad (8.4)$$

where we choose the square root in such a way that $\varepsilon \neq 0$, which is possible because of our assumption that $\Delta \neq 0$. If we choose a third root $u \in K$ of ε and put $v := -\frac{p}{u}$, the famous formulas of Cardano ([Bos93, 6.1]) imply $f = (X - x_1)(X - x_2)(X - x_3)$ with

$$x_1 = u + v,$$
$$x_2 = \zeta u + \zeta^2 v,$$
$$x_3 = \zeta^2 u + \zeta v,$$

therefore establishing $M' \subseteq L(\zeta, u)$. Then $\det \begin{pmatrix} 1 & 1 \\ \zeta & \zeta^2 \end{pmatrix} = \zeta(\zeta - 1) \neq 0$ shows we can express u and v in terms of x_1 and x_2 which proves the other inclusion $L(\zeta, u) \subseteq k'(x_1, x_2) = M'$.

(2) We set $k' := k(\zeta)$, $L' := L(\zeta)$ and get the following Hasse diagram:

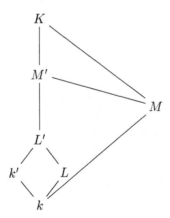

Let $\mu_3 \subset k'^{\times}$ denote the group of third roots of unity. Then the inflation-restriction-sequences associated to the towers $K/M'/L$, $K/M'/L'$, $M'/L'/L$ and $K/L'/L$ induce the following commutative diagram of abelian groups with exact rows and columns:

$$
\begin{array}{ccc}
1 & & 1 \\
\downarrow & & \downarrow \\
H^1(L'/L,\mu_3) & \!\!=\!\!\!=\!\! & H^1(L'/L,\mu_3) \\
\downarrow {\scriptstyle \mathrm{inf}} & & \downarrow {\scriptstyle \mathrm{inf}} \\
\end{array}
$$

$$
\begin{array}{ccccccc}
1 \longrightarrow & H^1(M'/L,\mu_3) & \xrightarrow{\;\mathrm{inf}\;} & H^1(L,\mu_3) & \xrightarrow{\;\mathrm{res}\;} & H^1(M',\mu_3) \\
& \downarrow {\scriptstyle \mathrm{res}} & & \downarrow {\scriptstyle \mathrm{res}} & & \| \\
1 \longrightarrow & H^1(M'/L',\mu_3) & \xrightarrow{\;\mathrm{inf}\;} & H^1(L',\mu_3) & \xrightarrow{\;\mathrm{res}\;} & H^1(M',\mu_3)
\end{array}
$$

Because of $[L' : L] \le 2$ and $(2,3) = 1$, we have $H^1(L'/L,\mu_3) = 1$, and if we use this fact and the Kummer-isomorphism for μ_3, we can update the diagram as follows:

$$
\begin{array}{ccccccc}
& 1 & & 1 & & \\
& \downarrow & & \downarrow & & \\
1 \longrightarrow & H^1(M'/L,\mu_3) & \xrightarrow{\;\mathrm{inf}\;} & L^\times/L^{\times 3} & \xrightarrow{\;\mathrm{res}\;} & M'^\times/M'^{\times 3} \\
& \downarrow {\scriptstyle \mathrm{res}} & & \downarrow {\scriptstyle \mathrm{res}} & & \| \\
1 \longrightarrow & H^1(M'/L',\mu_3) & \xrightarrow{\;\mathrm{inf}\;} & L'^\times/L'^{\times 3} & \xrightarrow{\;\mathrm{res}\;} & M'^\times/M'^{\times 3}
\end{array}
$$

Consider the elements $[\varepsilon]$ and $[\varepsilon']$ of $L^\times/L^{\times 3}$. By assumption, both are third powers in M', so their restriction to $M'^\times/M'^{\times 3}$ is zero, and consequently they must be the inflation of elements of $H^1(M'/L,\mu_3)$ which we want to denote by $\tilde\varepsilon$ and $\tilde\varepsilon'$.

First consider the case where $\tilde\varepsilon$ or $\tilde\varepsilon'$ is the trivial class — without loss of generality we can assume that it is $\tilde\varepsilon$. Then $[\varepsilon]$ must be trivial, too, so that ε is a third power in L and hence in L'. But L' contains μ_3, so that all three third roots of ε, namely u, ζu and $\zeta^2 u$ must lie in L'. Then $M' = L'(u)$ implies $L' = M'$, and because of $M' = L'(u')$ we have $u' \in L'$, so that ε' is a third power in L'. Since we already know that the restriction $L^\times/L^{\times 3} \to L'^\times/L'^{\times 3}$ is injective, it follows that ε' is also a third power in L, so $[\varepsilon] = [1] = [\varepsilon']$ in $L^\times/L^{\times 3}$, which means that there is indeed an $\eta \in L^\times$ with $\varepsilon' = \varepsilon\eta^3$, thus proving part (2) of the lemma in this case.

Let us now consider the other case where neither $\tilde{\varepsilon}$ nor $\tilde{\varepsilon}'$ are trivial. If we had $\tilde{\varepsilon} = \tilde{\varepsilon}'$ (and thus $[\varepsilon] = [\varepsilon']$), this would again be equivalent to the existence of an $\eta \in L^\times$ with $\varepsilon' = \varepsilon\eta^3$, and we would be done. So we only have to consider the case where $\tilde{\varepsilon}$ and $\tilde{\varepsilon}'$ are *distinct, non-trivial* classes.

Because of $\mu_3 \subset L'$, the Galois module μ_3 is endowed with the trivial action of $G_{M'/L'} := \mathrm{Gal}(M'/L')$, so we have

$$H^1(M'/L', \mu_3) = \mathrm{Hom}_{\mathrm{Groups}}(G_{M'/L'}, \mu_3). \tag{8.5}$$

Furthermore, by hypothesis M' is the splitting field of the polynomial $X^3 - \varepsilon \in L'[X]$ which is either irreducible or the product of three linear terms since $\mu_3 \subset L'$. Therefore the degree of M' over L' must be either one or three.

If this degree was one (and thus $M' = L'$), we would have $H^1(M'/L', \mu_3) = 1$ and then also $H^1(M'/L, \mu_3) = 1$ because of the injectivity of the restriction map. But then it would follow that $\tilde{\varepsilon} = \tilde{\varepsilon}' = 1$ — a contradiction.

Therefore we must have $[M' : L'] = 3$, so that $G_{M'/L'}$ is cyclic of order three. Then (8.5) implies that $H^1(M'/L', \mu_3)$ is also cyclic of order three, and our assumptions on $\tilde{\varepsilon}$ and $\tilde{\varepsilon}'$ imply $\tilde{\varepsilon}' = \tilde{\varepsilon}^2$. In particular, the quotient $\frac{\varepsilon'}{\varepsilon^2}$ is a third power $\tilde{\eta}^3$ in L, and we get

$$\frac{1}{\varepsilon'} = \frac{1}{\varepsilon^2\tilde{\eta}^3} = \varepsilon \cdot \left(\frac{1}{\varepsilon\tilde{\eta}}\right)^3$$

as claimed.

Finally, assume that δ is not a square in $k^\times$. We then have to show that $\frac{\tau(\varepsilon')}{\varepsilon} = \frac{\tau(\varepsilon^2\tilde{\eta}^3)}{\varepsilon}$ is a third power in L. First we claim that this is equivalent to showing that $\tau(\varepsilon) \cdot \varepsilon$ is a third power in L:

$$\frac{\tau(\varepsilon^2\tilde{\eta}^3)}{\varepsilon} = \eta^3 \Leftrightarrow \frac{\varepsilon^2\tilde{\eta}^3}{\tau(\varepsilon)} = \left(\tau(\eta)\right)^3 \Leftrightarrow \frac{\tau(\varepsilon)}{\varepsilon^2\tilde{\eta}^3} = \left(\frac{1}{\tau(\eta)}\right)^3$$

$$\Leftrightarrow \tau(\varepsilon) \cdot \varepsilon = \left(\frac{\varepsilon\tilde{\eta}}{\tau(\eta)}\right)^3.$$

Using the explicit formula (8.4) and noting that $\tau(\sqrt{\delta}) = -\sqrt{\delta}$, since δ

is not a square in k, we get:

$$\tau(\varepsilon) \cdot \varepsilon = \left(-q - \frac{1}{2}\sqrt{\delta}\right)\left(-q + \frac{1}{2}\sqrt{\delta}\right) = q^2 - \frac{\delta}{4}$$

$$= q^2 - (p^3 + q^2) = -p^3 = (-p)^3 \in k^{\times 3} \subset L^{\times 3}.$$

This completes the proof of the lemma.

$\square$

Lemma 8.4   *Using the notation of theorem 8.1, let M be the splitting field of $f := \frac{1}{a} \cdot Q(X, 1)$ as in lemma 8.2, let M' be the field $M(\zeta)$ as in lemma 8.3, and put $\varepsilon' := e \cdot \sqrt{\delta}$. Then ε' is the third power of an element $u' \in M'$, and we have $M' = k(\sqrt{\delta}, \zeta, u')$.*

Proof.   Let g be the monic polynomial

$$g(X) := \frac{1}{a} \cdot f\left(X - \frac{b}{3a}\right)$$

$$= X^3 + 3\underbrace{\left(\frac{3ac - b^2}{9a^2}\right)}_{=:p} X + 2\underbrace{\left(\frac{27a^2 d + 2b^3 - 9abc}{54a^3}\right)}_{=:q} \in k[X].$$

If we denote the discriminants of f and g by Δ_f respectively Δ_g, it follows from lemma 8.1(2) that

$$\Delta(Q) = a^4 \cdot \Delta_f = a^4 \cdot \Delta_g = -a^4 \cdot 4 \cdot 27 \cdot (p^3 + q^2),$$

$$\sqrt{\delta} \;= \sqrt{-\frac{\Delta(Q)}{27}} = \sqrt{4a^4(p^3 + q^2)} = 2a^2\sqrt{p^3 + q^2}.$$

Using this we get:

$$\varepsilon' = \left(\frac{a}{2} - \frac{27a^2 d + 2b^3 - 9abc}{2\Delta(Q)}\sqrt{\delta}\right) \cdot \sqrt{\delta}$$

$$= \frac{a}{2}\sqrt{\delta} - \frac{54a^3 q \cdot 4a^4(p^3 + q^2)}{-2a^4 \cdot 4 \cdot 27 \cdot (p^3 + q^2)}$$

$$= a^3\sqrt{p^3 + q^2} + a^3 q$$

$$= a^3 \cdot (q + \sqrt{p^3 + q^2})$$

$$= (-a)^3 \cdot (-q - \sqrt{p^3 + q^2}).$$

If we now take a look at the proof of lemma 8.3(1), we see that between the element ε constructed there and our element ε', there is the relation $\varepsilon' = (-a)^3 \cdot \varepsilon$. And because ε has the desired properties, the same must be true for ε', and the lemma follows.   □

Proof. *(of theorem 8.1)*

First we want to show that the class (L_δ, e) respectively $(L_\delta, (e, \frac{\sqrt{\delta}}{e}))$ in $H^1_{\text{cont}}(G, \mu_3 \wr S_2)$ only depends on the isomorphism class $[Q]$ of Q in $\mathcal{F}$. To this end, we first note that the image of δ in $k^\times/k^{\times 2}$ (and thus the isomorphism class of L_δ) indeed only depends on $[Q]$ because of lemma 8.1.

Now let $Q'(X, Y) = a'X^3 + b'X^2Y + c'XY^2 + d'Y^3$ be another member of the class $[Q]$ with $a' \neq 0$, and let $\delta' := \delta(Q')$ and $e' := e(Q')$ be its invariants. According to corollary 7.4 we have to prove that there is an automorphism $a \in A_\delta$ and an element $y \in L_\delta^\times$ with $e' = a[ey^3]$ respectively $(e', \frac{\sqrt{\delta'}}{e'}) = a[(e, \frac{\sqrt{\delta}}{e})y^3]$.

Use the notation of lemma 8.3, and first consider the case where δ is *not* a square in k. According to lemma 8.1(1), the quotient $\frac{\delta'}{\delta}$ is a sixth power in $k^\times$, so we have $\frac{\delta'}{\delta} = \nu^6$ for a suitable $\nu \in k^\times$. so by replacing ν with $(-\nu)$ if necessary, we get $\frac{\sqrt{\delta'}}{\sqrt{\delta}} = \nu^3$. From lemma 8.3 and lemma 8.4 we know that there is an $\eta \in L(\sqrt{\delta})^\times = L_\delta^\times$ with $\sqrt{\delta'}e' = \sqrt{\delta}e\eta^3$ or $\tau(\sqrt{\delta'}e') = \sqrt{\delta}e\eta^3$ (and hence $e' = e\eta^3 \frac{\sqrt{\delta}}{\sqrt{\delta'}}$ or $e' = \frac{\tau[\sqrt{\delta}e\eta^3]}{\sqrt{\delta'}}$). If we put $a := \text{id}$ and $y := \frac{\eta}{\nu}$ in the first case and $a := \tau$ and $y := -\frac{\eta}{\nu}$ in the second case, we get (note that $\tau(\sqrt{\delta}) = -\sqrt{\delta}$ and $\tau(\nu) = \nu$ and therefore $\tau(\sqrt{\delta'}) = -\sqrt{\delta'}$):

<u>1. case:</u> $a[ey^3] = e\left(\dfrac{\eta}{\nu}\right)^3 = e\eta^3 \cdot \dfrac{\sqrt{\delta}}{\sqrt{\delta'}} = e'$,

<u>2. case:</u> $a[ey^3] = \tau\left[e\left(-\dfrac{\eta}{\nu}\right)^3\right] = \tau\left[e\eta^3 \cdot \left(-\dfrac{\sqrt{\delta}}{\sqrt{\delta'}}\right)\right] = \dfrac{\tau\left[\sqrt{\delta}e\eta^3\right]}{\sqrt{\delta'}} = e'$.

Now assume that δ *is* a square in k. Then there is — again according to lemma 8.3 and lemma 8.4 — an $\eta \in k^\times$ with $\sqrt{\delta'}e' = \sqrt{\delta}e\eta^3$ or $\frac{1}{\sqrt{\delta'}e'} = \sqrt{\delta}e\eta^3$. Put $a := \text{id}$ and $y := (\frac{\eta}{\nu}, \frac{\nu^2}{\eta})$ in the first case and $a := \tau$ and

$y := (\frac{\sqrt{\delta'}\eta}{\nu}, \frac{\nu^2}{\sqrt{\delta'}\eta})$ in the second case, and we get:

<u>1. case:</u> $\quad a\left[\left(e, \frac{\sqrt{\delta}}{e}\right)y^3\right] = \left(e\eta^3 \cdot \frac{\sqrt{\delta}}{\sqrt{\delta'}}, \sqrt{\delta'} \cdot \frac{\sqrt{\delta'}}{\sqrt{\delta}} \cdot \frac{1}{e\eta^3}\right) = \left(e', \frac{\sqrt{\delta'}}{e'}\right),$

<u>2. case:</u> $\quad a\left[\left(e, \frac{\sqrt{\delta}}{e}\right)y^3\right] = \tau\left(\delta'\sqrt{\delta}e\eta^3, \frac{1}{\sqrt{\delta\delta'}e\eta^3}\right) = \left(e', \frac{\sqrt{\delta'}}{e'}\right).$

With this we have shown that the isomorphism class of the twisted Fermat equation $P_2^3\{(L_\delta, e)\}$ respectively $P_2^3\{(L_\delta, (e, \sqrt{\delta}/e))\}$ associated to Q only depends on the isomorphism class $[Q]$. It remains to be shown that this twisted Fermat equation is really *isomorphic* to Q. But we already know from corollary 8.1 that there indeed is a twisted Fermat equation $P :=$ $P_2^3\{(L_\delta', x)\}$ which is isomorphic to Q, so using what we have shown above, it is enough to prove that the twisted Fermat equation associated to P is isomorphic to P.

First consider the case $\delta \notin k^{\times 2}$ and $x = y + z\alpha$ with $\alpha^2 = \delta'$ and $y, z \in k$. Then using the k-basis $\{1, \alpha\}$ of $L_{\delta'}$, because of proposition 7.3 we get:

$$
\begin{aligned}
P &= P_2^3\{(\delta', x)\} \\
&= \mathrm{Tr}_{L_{\delta'}/k}\left[(y + z\alpha) \cdot (X + \alpha Y)^3\right] \\
&= \mathrm{Tr}_{L_{\delta'}/k}\Big[(y + z\alpha)X^3 + (3y\alpha + 3\delta'z)X^2Y \\
&\qquad\qquad + (3\delta'y + 3\delta'z\alpha)XY^2 + (\delta'^2z + \delta'y\alpha)Y^3\Big] \\
&= (2y)X^3 + (6\delta'z)X^2Y + (6\delta'y)XY^2 + (2\delta'^2z)Y^3,
\end{aligned}
$$

so we have $a = 2y$, $b = 6\delta'z$, $c = 6\delta'y$ and $d = 2\delta'^2z$.

<u>1. case:</u> $y \neq 0$.
Then $a = 2y \neq 0$, and we can immediately compute the associated invariants:

$$\Delta(P) = -1728\,\delta'^3(y^2 - \delta'z^2)^2 = (-27\delta') \cdot \left[8\delta'(y^2 - \delta'z^2)\right]^2,$$

$$\delta(P) = \delta' \cdot \left[8\delta'(y^2 - \delta'z^2)\right]^2,$$

$$e(P) = y - \frac{\delta'^2(y^2 - \delta'z^2)z}{\delta'^2(y^2 - \delta'z^2)} \cdot \alpha = y - z\alpha = \tau[y + z\alpha] = \tau[x].$$

<u>2. case:</u> $y = 0$.
Because of $x \neq 0$, we then must have $z \neq 0$, and we can consider P_A for

$A := \left(\begin{smallmatrix} 0 & 1 \\ 1 & 0 \end{smallmatrix}\right):$

$$P_A = (2\delta'^2 z)X^3 + (6\delta' y)X^2 Y + (6\delta' z)XY^2 + (2y)Y^3.$$

Since $\det(A) = -1$, we know from lemma 8.1(1) that $\Delta(P_A) = \Delta(P)$, so that we get the same δ as above. For e we compute

$$e = \delta'^2 z - \frac{\delta'^4 z^2 y - \delta'^3 y^3}{\delta'^2 (y^2 - \delta_1 z^2)} \cdot \alpha = \delta'^2 z + \delta' y\alpha = (y + \alpha z) \cdot \alpha^3 = x \cdot \alpha^3.$$

— we see that the theorem follows in both cases.

It remains the case where δ' is a square in k. Let $x = (y, z)$ with $y, z \in k^\times$, then using the k-basis $\{(1,0), (0,1)\}$ of $L_{\delta'} = k \times k$ we get from proposition 7.3:

$$P = P_2^3\{(\delta', x)\} = \text{Tr}_{(k \times k)/k}\left((y, z) \cdot [(X, 0) + (0, Y)]^3\right)$$
$$= \text{Tr}_{(k \times k)/k}\left(yX^3, zY^3\right) = yX^3 + zY^3,$$

and therefore have $a = y$, $b = c = 0$ and $d = z$. Fortunately we do not have to distinguish between two cases this time because of $y, z \in k^\times$. We compute:

$$\Delta(P) = -27y^2 z^2,$$
$$\delta = y^2 z^2 = (yz)^2,$$
$$e = \frac{y}{2} - \frac{27y^2 z}{-54y^2 z^2} \cdot (\pm yz) = \frac{y}{2} \pm \frac{y}{2} = y,$$

since we are supposed to choose the squareroot of $y^2 z^2$ that gives us an $e \neq 0$. Then

$$(e, \frac{\sqrt{\delta}}{e}) = (y, \frac{yz}{y}) = (y, z) = x,$$

and the statement is shown in this case, too. Thus the proof of the theorem is complete. □

Example 8.1 Consider the binary cubic equation $Q(X, Y) := X^2 Y + XY^2$ over $\mathbb{Q}$. First, we compute its discriminant and get

$$\Delta(Q) = 0 + 0 + 1^2 \cdot 1^2 - 0 - 0 = 1.$$

In particular, Q is non-singular, and we can ask: "To which equation $P_2^3\{L, x\}$ is Q isomorphic?"

Unfortunately, the coefficient of X^3 is zero, so that we first have to transform Q by means of lemma 8.2. Using the notation of lemma 8.2, we must find a $\rho \in \mathbb{Q}$ with $P(\rho) := \rho^2 + \rho \neq 0$ — we can for example take $\rho := 1$. Then $A = \left(\begin{smallmatrix} 1 & 0 \\ 1 & 1 \end{smallmatrix}\right)$, and we get the isomorphic equation

$$Q_A(X, Y) = Q(X, X + Y) = 2X^3 + 3X^2Y + XY^2,$$

i.e. $a = 2$, $b = 3$, $c = 1$ and $d = 0$.

Now we can compute the invariants δ and e (here we can use the fact that according to lemma 8.1(1), the discriminants of Q and Q_A are equal):

$$\delta = -\frac{1}{27},$$

$$e = \frac{2}{2} - \frac{0 + 2 \cdot 3^3 - 9 \cdot 2 \cdot 3 \cdot 1}{2 \cdot 1} \sqrt{-\frac{1}{27}} = 1.$$

So we have $L_\delta = \mathbb{Q}\left(\sqrt{-\frac{1}{27}}\right) = \mathbb{Q}(\sqrt{-3})$, and we get the final result

$$\boxed{Q \cong P_2^3\left\{\left(\mathbb{Q}(\sqrt{-3}), 1\right)\right\}}.$$

Chapter 9

Forms of the Fermat equation II

As in chapter 7, let r and n be positive natural numbers, let k be a field whose characteristic is greater than r, let $K := \bar{k}$ be a separable closure of k, let $G := \mathrm{Gal}(K/k)$ be the absolute Galois group of k, and let P_n^r be the Fermat equation $X_1^r + \ldots + X_n^r \in k[X_i]$.

In chapter 7, we studied P_n^r as an object of $\mathcal{F}_k^{n,r}$, but in this chapter we primarily want to study it as an object of $\widetilde{\mathcal{F}_k^{n,r}}$ and try to determine its $\widetilde{\mathcal{F}_K^{n,r}} / \widetilde{\mathcal{F}_k^{n,r}}$-forms.

Proposition 9.1 *The diagonal embedding $\mu_r \hookrightarrow \mu_r \wr S_n$, which sends $\zeta \in \mu_r$ to $(\zeta, \ldots, \zeta) \cdot 1_{S_n}$, identifies μ_r with a central (and thus normal) subgroup of the wreath product $\mu_r \wr S_n$ — denote the quotient by $\tilde{A} := \tilde{A}_n^r$:*

$$1 \longrightarrow \mu_r \longrightarrow \mu_r \wr S_n \longrightarrow \tilde{A} \longrightarrow 1.$$

The embedding $\mu_r \wr S_n \hookrightarrow GL(n, K)$ defined in lemma 7.1 induces an embedding $\tilde{A} \hookrightarrow PGL(n, K)$ which factorizes over the subgroup $A(P_n^r)$ of $PGL(n, K)$, the automorphism group of P_n^r in $\widetilde{\mathcal{F}_K^{n,r}}$:

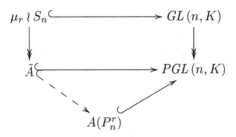

If $r \geq 3$, we even get an isomorphism $A(P_n^r) \cong \tilde{A}$.

153

Proof.

- *central subgroup:* Trivial.
- $\tilde{A} \to PGL(n, K)$ *well defined:* Obvious, because the elements of μ_r are mapped to elements of $K^\times$ in $GL(n, K)$ and thus to the unit element in $PGL(n, K)$.
- $\tilde{A} \to PGL(n, K)$ *injective:* Let $\overline{(\zeta_i)_i \cdot \sigma}$ be an element of $\tilde{A}$ which is mapped to the unit element of $PGL(n, K)$. Then the element $(\zeta_i)_i \cdot \sigma$ of $\mu_r \wr S_n$ is mapped to a scalar $c \in K^\times$ in $GL(n, K)$, which immediately implies that $\sigma = 1_{S_n}$ and $\zeta_1 = \ldots = \zeta_n = c$. Therefore $\overline{(\zeta_i)_i \cdot \sigma}$ is the unit element of $\tilde{A}$.
- $\tilde{A}$ *subgroup of* $A(P_n^r)$: As stated in the proof of proposition 7.1, all elements of $\mu_r \wr S_n$ are contained in the subgroup $\mathrm{Aut}_{\mathcal{F}_K^{n,r}}(P_n^r)$ of $GL(n, K)$ and are hence mapped to elements of $\mathrm{Aut}_{\widetilde{\mathcal{F}_K^{n,r}}}(P_n^r)$ by definition of morphisms in $\widetilde{\mathcal{F}_K^{n,r}}$. By construction of the map $\tilde{A} \hookrightarrow A(P_n^r)$, the claim follows from this.
- *isomorphism in case* $r \geq 3$: So let $r \geq 3$, and let $\bar{B} \in PGL(n, K)$ be an $\widetilde{\mathcal{F}_K^{n,r}}$-automorphism of P_n^r, represented by $B \in GL(n, K)$. Then by definition of morphisms in $\widetilde{\mathcal{F}_K^{n,r}}$, there is an element $c \in K^\times$ with $P_n^r(BX) = c \cdot P_n^r$. Let $d \in K^\times$ be an r-th root of c (remember that the characteristic of k is greater than r), and let $B' \in GL(n, K)$ be the matrix $\frac{1}{d}B$. Then

$$P_n^r(B'X) = P_n^r(\tfrac{1}{d}BX) = \left(\tfrac{1}{d}\right)^r \cdot P_n^r(BX) = P_n^r,$$

so B' is an $\mathcal{F}_K^{n,r}$-automorphism of P_n^r. According to proposition 7.1, the matrix B' therefore is an element of $\mu_r \wr S_n$, and the claim follows, since obviously $\bar{B}' = \bar{B}$ in $PGL(n, K)$.

□

Lemma 9.1 *When μ_r is embedded diagonally into μ_r^n via $\zeta \mapsto (\zeta, \ldots, \zeta)$, the natural left-action of S_n on μ_r^n leaves the subgroup μ_r invariant, so that we get a short exact sequence of abelian S_n-groups*

$$1 \longrightarrow \mu_r \longrightarrow \mu_r^n \longrightarrow \mu_r^n/\mu_r \longrightarrow 1. \tag{9.1}$$

We can therefore consider the group $(\mu_r^n/\mu_r) \rtimes S_n$, and we get a canonical *isomorphism of groups*

$$\tilde{A} = (\mu_r^n \rtimes S_n)/\mu_r \longrightarrow (\mu_r^n/\mu_r) \rtimes S_n$$

$$\overline{(\zeta_i)_i \cdot \sigma} \quad \mapsto \quad \overline{(\zeta_i)_i} \cdot \sigma.$$

From now on, we want to identify $\tilde{A}$ with $(\mu_r^n/\mu_r) \rtimes S_n$ via this isomorphism.

Proof. It is clear that μ_r is an S_n-invariant subgroup of μ_r^n, because S_n fixes elements of the form $(\zeta, \ldots, \zeta)$, so we get sequence (9.1) from lemma 4.6. It is then trivial to check that the map $\tilde{A} \to (\mu_r^n/\mu_r) \rtimes S_n$ is a well defined group isomorphism. $\square$

Lemma 9.2 *Because of lemma 9.1, we can define a continuous left-G-action on $\tilde{A}$ as follows:*

$$G \times \left[(\mu_r^n/\mu_r) \rtimes S_n \right] \longrightarrow (\mu_r^n/\mu_r) \rtimes S_n \tag{9.2}$$

$$(s, \quad \overline{(\zeta_i)_i} \cdot \sigma) \quad \mapsto \quad \overline{({}^s\zeta_i)_i} \cdot \sigma.$$

With respect to this action, the normal subgroup $(\mu_r^n/\mu_r)G$- of $\tilde{A}$ is G-invariant, and if $r \geq 3$, then the isomorphism $\tilde{A} \cong \operatorname{Aut}_{\widetilde{\mathcal{F}_K^{n,r}}}(P_n^r)$ from proposition 9.1 is an isomorphism of discrete G-groups (remember that $\operatorname{Aut}_{\widetilde{\mathcal{F}_K^{n,r}}}(P_n^r)$ is a G-group because of example 3.3(4) and a discrete G-group because of example 3.5(2)).

Proof.

- *continuous left-G-action:* The group μ_r is a discrete G-group, so by lemma 4.8, the wreath product $\mu_r \wr S_n$ is one as well, and it is clear that the normal subgroup μ_r of $\mu_r \wr S_n$ is G-invariant. So according to lemma 4.6, the quotient $\tilde{A}$ is also a discrete G-group, and it is obvious that the so defined G-action on $\tilde{A}$ is given by (9.2) if we identify $\tilde{A}$ with $(\mu_r^n/\mu_r) \rtimes S_n$ according to lemma 9.1.
- *μ_r^n/μ_r is G-invariant:* The G-action thus defined on $\tilde{A}$ is the same that we would get if we applied corollary 4.2 to $S := G$, $T := S_n$ and $A := \mu_r^n/\mu_r$, so it follows in particular that μ_r^n/μ_r is a G-invariant normal subgroup of $\tilde{A}$, a fact that can of course also be deduced from (9.2) directly.

- *isomorphism for $r \geq 3$:* By definition of a group action, any G-action on $\tilde{A} \cong \mathrm{Aut}_{\widetilde{\mathcal{F}_K^{n,r}}}(P_n^r)$ must satisfy $^s\overline{((\zeta_i)_i \cdot \sigma)} = \overline{^s(\zeta_i)_i} \cdot {}^s\sigma$, and according to example 3.3(4), G acts on an element of $\mathrm{Aut}_{\widetilde{\mathcal{F}_K^{n,r}}}(P_n^r) \leq \mathrm{PGL}\,(n,K)$ by acting on the entries of a representing matrix. This implies $\overline{^s(\zeta_i)_i} = \overline{(^s\zeta_i)_i}$ and $^s\sigma = \sigma$, because σ is represented by a permutation matrix in $\mathrm{GL}\,(n,K)$ with entries in $\{0,1\}$ and is thus fix under the G-action. So we see that the natural action on $\mathrm{Aut}_{\widetilde{\mathcal{F}_K^{n,r}}}(P_n^r)$ is just the action on $\tilde{A}$ defined by (9.2). □

Lemma 9.3 *From now on let $p: \mu_r \wr S_n \twoheadrightarrow \tilde{A}$ denote the natural projection. Then the following diagram of discrete G-groups commutes and has exact rows and columns:*

(9.3)

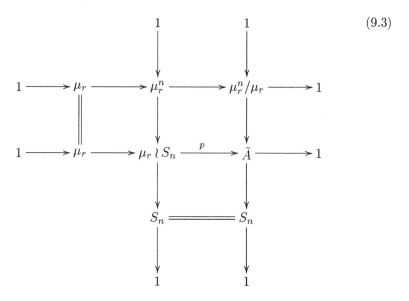

Proof. This is obvious. □

Lemma 9.4 *Consider the morphism of coefficient extensions from example 3.4(2). According to example 3.5(2) and example 3.5(3), both coefficient extensions involved are continuous, so corollary 6.1 shows that we get the* *following commuting diagram in the category of pointed sets (note that the left vertical morphism is an isomorphism because of corollary 7.6):*

$$
\begin{array}{ccc}
E(\mathcal{F}_K^{n,r}/\mathcal{F}_k^{n,r}, P_n^r) & \xrightarrow[\;[Q]\mapsto[Q]\;]{\varphi} & E(\widetilde{\mathcal{F}_K^{n,r}}/\widetilde{\mathcal{F}_k^{n,r}}, P_n^r) \\
\Big\downarrow{\scriptstyle\vartheta}\;\wr & & \Big\downarrow{\scriptstyle\vartheta} \\
H_{cont}^1(G, Aut_{\mathcal{F}_K^{n,r}}(P_n^r)) & \xrightarrow[\psi]{(a_s)\mapsto(\bar{a}_s)} & H_{cont}^1(G, A(P_n^r)).
\end{array}
$$

$$(9.4)$$

Furthermore, the morphism φ is surjective.

Proof. We only have to prove that φ is surjective: Let Q be an arbitrary $\widetilde{\mathcal{F}_K^{n,r}}/\widetilde{\mathcal{F}_k^{n,r}}$-form of P_n^r. So there exists an isomorphism $\bar{A} \in \mathrm{PGL}(n,K)$ from Y to P_n^r in $\widetilde{\mathcal{F}_K^{n,r}}$. Let $A \in \mathrm{GL}(n,K)$ be a matrix representing $\bar{A}$. By definition of morphisms in $\widetilde{\mathcal{F}_K^{n,r}}$, there then is an element $\lambda \in K^\times$ with $P_n^r(AX) = \lambda \cdot Q$.

Therefore if we put $A' := \frac{1}{\sqrt[r]{\lambda}} \cdot A$, we get

$$
P_n^r(A'X) = \left(\frac{1}{\sqrt[r]{\lambda}}\right)^r \cdot P_n^r(AX) = \frac{1}{r} \cdot r \cdot Q = Q,
$$

which means that A' defines an isomorphism from Q to P_n^r in $\mathcal{F}_K^{n,r}$. This shows that Q is also an $\mathcal{F}_K^{n,r}/\mathcal{F}_k^{n,r}$-form of P_n^r and thus proves the surjectivity of φ. $\qquad\square$

Lemma 9.5 *Let*

$$
\begin{array}{ccc}
A & \xrightarrow{\alpha} & B \\
{\scriptstyle\beta}\Big\downarrow & & \Big\downarrow{\scriptstyle\gamma} \\
C & \xrightarrow[\delta]{} & D
\end{array}
$$

be a commutative diagram in the category of (pointed) sets. Then the images of γ and δ coincide in D.

Proof.   This is obvious: If d is an arbitrary element of D, then

$$d \in \text{Im}\,(\delta) \iff \exists c \in C : d = \delta(c)$$
$$\overset{\beta \text{ surjective}}{\iff} \exists a \in A : d = \delta(\beta(a)) = \gamma(\alpha(a))$$
$$\overset{\alpha \text{ surjective}}{\iff} \exists b \in B : d = \gamma(b) \iff d \in \text{Im}\,(\gamma)$$
$$\square$$

Corollary 9.1   *If $r \geq 3$, then the image of $E(\widetilde{\mathcal{F}_K^{n,r}}/\widetilde{\mathcal{F}_k^{n,r}}, P_n^r)$ under ϑ equals the image of $H_{cont}^1(G, Aut_{\mathcal{F}_k^{n,r}}(P_n^r))$ under ψ, so that we have an isomorphism of pointed sets*

$$\boxed{E(\widetilde{\mathcal{F}_K^{n,r}}/\widetilde{\mathcal{F}_k^{n,r}}, P_n^r) \xrightarrow{\sim} Im\left(H_{cont}^1(G, \mu_r \wr S_n) \xrightarrow{H_{cont}^1(p)} H_{cont}^1(G, \tilde{A}) \right)}$$

Proof.   From lemma 9.5, applied to diagram (9.4), we know that the images of ϑ and ψ coincide in $H_{\text{cont}}^1(G, A(P_n^r))$. Now consider the following commuting diagram of discrete G-groups (where the vertical isomorphisms are those from lemma 7.2 and lemma 9.2):

$$
\begin{array}{ccc}
\mu_r \wr S_n & \xrightarrow{\ \ p\ \ } & \tilde{A} \\
{\wr}\big\downarrow & & \big\downarrow{\wr} \\
\text{Aut}_{\mathcal{F}_K^{n,r}}(P_n^r) & \xrightarrow[A \mapsto \tilde{A}]{} & A(P_n^r).
\end{array}
$$

We therefore get the following induced commutative diagram of pointed sets:

$$
\begin{array}{ccc}
H_{\text{cont}}^1(G, \mu_r \wr S_n) & \xrightarrow{H_{\text{cont}}^1(p)} & H_{\text{cont}}^1(G, \tilde{A}) \\
{\wr}\big\downarrow & & \big\downarrow{\wr} \\
H_{\text{cont}}^1(\text{Aut}_{\mathcal{F}_K^{n,r}}(P_n^r)) & \xrightarrow[\psi]{} & H_{\text{cont}}^1(G, A(P_n^r)),
\end{array}
$$

and thus an isomorphism of pointed sets

$$\text{Im}\,(\psi) \cong \text{Im}\left(H_{\text{cont}}^1(G, \mu_r \wr S_n) \xrightarrow{H_{\text{cont}}^1(p)} H_{\text{cont}}^1(G, \tilde{A}) \right).$$

Combining this with the equation $\operatorname{Im}(\vartheta) = \operatorname{Im}(\psi)$ proven above and the fact $\operatorname{Im}(\vartheta) \cong E(\widetilde{\mathcal{F}_K^{n,r}}/\widetilde{\mathcal{F}_k^{n,r}}, P_n^r)$ (which follows from ϑ being injective), the proof is completed. $\qquad\qquad\square$

Proposition 9.2 *Two classes in $H^1_{cont}(G, \mu_r \wr S_n)$, according to corollary 7.4 given by pairs (L, x) and (L', x'), are mapped to the same class in $H^1_{cont}(G, \tilde{A})$ under $H^1_{cont}(p)$ if and only if there is an element $\lambda \in k^\times$ with $(L', x') = (L, \lambda x)$ in $H^1_{cont}(G, \mu_r \wr S_n)$.*

Proof. We denote the 1-cocycle of G in $\mu_r \wr S_n$ defined by (L, x) by $b = (b_s)$ and twist the lower exact row of diagram (9.3) with this cocycle to get the following short exact sequence of discrete G-groups:

$$1 \to \mu_{r,b} \to (\mu_r \wr S_n)_b \xrightarrow{p} \tilde{A}_b \to 1. \tag{9.5}$$

Taking the long exact sequence associated to sequence (9.5), we get the following exact sequence of pointed sets:

$$H^1_{cont}(G, \mu_{r,b}) \longrightarrow H^1_{cont}(G, (\mu_r \wr S_n)_b) \xrightarrow{H^1_{cont}(p)} H^1_{cont}(G, \tilde{A}_b).$$

We know from proposition 9.1 that μ_r is a central subgroup of $\mu_r \wr S_n$, so that we have $\mu_{r,b} = \mu_r$ by lemma 4.11. Using the Kummer isomorphism (8.3), we therefore get an isomorphism of abelian groups

$$
\begin{aligned}
k^\times/(k^\times)^r &\xrightarrow{\sim} H^1_{cont}(G, \mu_{r,b}) \\
\lambda &\mapsto \left(\frac{s[\sqrt[r]{\lambda}]}{\sqrt[r]{\lambda}}\right)_s.
\end{aligned}
\tag{9.6}
$$

Now we apply corollary 4.6 to the short exact sequence

$$1 \to \mu_r \to \mu_r \wr S_n \xrightarrow{p} \tilde{A} \to 1$$

from (9.3) and to the element $b \in H^1_{cont}(G, \mu_r \wr S_n)$. It tells us that a class (L', x') of $H^1_{cont}(G, \mu_r \wr S_n)$ has the same image as b in $H^1_{cont}(G, \tilde{A})$ if and only if it is given by a cocycle of the form $(a_s \cdot b_s)_s$ for a cocycle (a_s) of G in $\mu_{r,b}$. According to (9.6) from above, this is the case if and only if (L', x') equals $\left(\frac{s[\sqrt[r]{\lambda}]}{\sqrt[r]{\lambda}} \cdot b_s\right)$ with $\lambda \in k^\times$. Substituting the explicit formula for b

from corollary 7.4 and denoting the 1-cocycle of G in S_n given by L by (c_s), we get:

$$\left(\frac{s\left[\sqrt[r]{\lambda}\right]}{\sqrt[r]{\lambda}} \cdot b_s\right) = \left(\left(\frac{s\left[\sqrt[r]{\lambda}\right] \cdot s\left[\sqrt[r]{s^{-1}\varphi x}\right]}{\sqrt[r]{\lambda} \cdot \sqrt[r]{\varphi x}}\right)_\varphi \cdot c_s\right)_s$$

$$= \left(\left(\frac{s\left[\sqrt[r]{s^{-1}\varphi(\lambda x)}\right]}{\sqrt[r]{\varphi(\lambda x)}}\right)_\varphi \cdot c_s\right)_s = (L, \lambda x).$$

(Note that all $s \in G$ and all $\varphi \in M$ are k-morphisms and hence map λ to itself.)

So we see that the class (L', x') has the same image as (L, x) in $H^1_{\text{cont}}(G, \tilde{A})$ if and only if it is of the form $(L, \lambda x)$ for a suitable $\lambda \in k^\times$, and this is exactly what was claimed in the proposition. $\qquad\qquad\square$

Corollary 9.2 *For an fcs-algebra L of degree n over k, we want to call two classes $[x]$ and $[x']$ in the set $Aut_k(L)\backslash(L^\times/L^{\times^r})$ equivalent if there is an element $\lambda \in k^\times$ such that $(L, x) = (L, \lambda x')$ in $H^1_{\text{cont}}(G, \mu_r \wr S_n)$. —
According to corollary 7.4, the fact $(L, x) = (L, \lambda x')$ is equivalent to the existence of elements $y \in L^\times$ and $a \in Aut_k(L)$ with $x = a[\lambda x' y^r]$.*

Being equivalent is an equivalence relation on $Aut_k(L)\backslash(L^\times/L^{\times^r})$, and we want to denote the set of equivalence classes by $Aut_k(L)\backslash(L^\times/L^{\times^r})/k^\times$. Then we have the following commuting diagram of pointed sets:

$$
\begin{array}{ccc}
\coprod\limits_{[L]} Aut_k(L))\backslash(L^\times/L^{\times^r}) & \xrightarrow{\quad can \quad} & \coprod\limits_{[L]} Aut_k(L))\backslash(L^\times/L^{\times^r})/k^\times \\
\downarrow{\wr} & & \uparrow \\
H^1_{\text{cont}}(G, \mu_r \wr S_n) & \xrightarrow{\quad H^1_{\text{cont}}(p) \quad} & H^1_{\text{cont}}(G, \tilde{A})
\end{array}
\qquad (9.7)
$$

Here the unions are taken over the set of isomorphism classes $[L]$ of fcs-algebras of degree n over k, and the special elements are represented by the pair $(k^n, 1)$.

Proof. First we want to show that being equivalent is really an equivalence relation: Reflexivity is trivial, so let us check symmetry: If $(L, x) \sim (L, x')$, then there are $\lambda \in k^\times$, $y \in L^\times$ and $a \in \mathrm{Aut}_k(L)$ with $x = a[\lambda x' y^r]$, so

$$x' = \frac{a^{-1}[x]}{\lambda y^r} = a^{-1} \left[\frac{1}{\lambda} \cdot x \cdot \left(\frac{1}{a[y]} \right)^r \right]$$

and hence $(L, x') \sim (L, x)$. Transitivity is also easy: If in addition to $(L, x) \sim (L, x')$ we have $(L, x') \sim (L, x'')$, then there are elements $\tilde{\lambda} \in k^\times$, $\tilde{y} \in L^\times$ and $\tilde{a} \in \mathrm{Aut}_k(L)$ with $x' = \tilde{a}[\tilde{\lambda} x'' \tilde{y}^r]$ and therefore

$$x = a[\lambda x' y^r] = a \left[\lambda \left(\tilde{a}[\tilde{\lambda} x'' \tilde{y}^r] \right) y^r \right] = (a\tilde{a}) \left[(\lambda\tilde{\lambda}) \cdot x'' \cdot \left(\tilde{y} \cdot \tilde{a}^{-1}[y] \right)^r \right],$$

so that $(L, x) \sim (L, x'')$.

Now according to proposition 9.2, the composition

$$\coprod_{[L]} \mathrm{Aut}_k(L)) \backslash (L^\times / L^{\times^r}) \xrightarrow{\sim} H^1_{\mathrm{cont}}(G, \mu_r \wr S_n) \xrightarrow{H^1_{\mathrm{cont}}(p)} H^1_{\mathrm{cont}}(G, \tilde{A})$$

factorizes over this equivalence relation, and the corollary follows. $\qquad \square$

Corollary 9.3 *Let $r \geq 3$. Then ϑ induces an isomorphism of pointed sets*

$$\boxed{E(\widetilde{\mathcal{F}_K^{n,r}}/\widetilde{\mathcal{F}_k^{n,r}}, P_n^r) \xrightarrow{\sim} \coprod_{[L]} Aut_k(L) \backslash (L^\times / L^{\times^r}) / k^\times}$$

The inverse is given by sending a pair (L, x) to the isomorphism class of the twisted equation $P_n^r\{(L, x)\}$.

Proof. Applying lemma 9.5 to diagram (9.7) gives us (since the right vertical map is injective):

$$\coprod_{[L]} \mathrm{Aut}_k(L)) \backslash (L^\times / L^{\times^r}) / k^\times$$

$$\xrightarrow{\sim} \mathrm{Im}\left(H^1_{\mathrm{cont}}(G, \mu_r \wr S_n) \xrightarrow{H^1_{\mathrm{cont}}(p)} H^1_{\mathrm{cont}}(G, \tilde{A}) \right),$$

and combining this with corollary 9.1, we get the desired isomorphism. It is then clear that this isomorphism is induced by ϑ and that its inverse is given as stated.   □

As a first application, we want to classify non-singular binary cubic equations over a finite field k in $\widetilde{\mathcal{F}}_k^{2,3}$:

Corollary 9.4   *Consider the special case where k is a finite field, and let Q be an arbitrary non-singular binary cubic equation over k. If we use the notation from corollary 8.2, then:*

(1) *If k contains the third roots of unity, then Q is isomorphic in $\widetilde{\mathcal{F}}_k^{2,3}$ to exactly one of the following three twisted Fermat equations:*

$$
\begin{aligned}
P_2^3\{(L_1,(1,1))\} &= X^3 + Y^3, \\
P_2^3\{(L_1,(1,\delta))\} &= X^3 + \delta Y^3, \\
P_2^3\{(L_\delta,1)\} &= 2X^3 + 6\delta XY^2.
\end{aligned}
$$

(2) *If k does not contain the third roots of unity, then Q is isomorphic in $\widetilde{\mathcal{F}}_k^{2,3}$ to exactly one of the following three twisted Fermat equations:*

$$
\begin{aligned}
P_2^3\{(L_1,(1,1))\} &= X^3 + Y^3, \\
P_2^3\{(L_\delta,1)\} &= 2X^3 + 6\delta XY^2 \\
P_2^3\{(L_\delta,\alpha)\} &= Tr_{k'/k}(\alpha) \cdot X^3 + 3 \cdot Tr_{k'/k}(\alpha\beta) \cdot X^2 Y \\
&\quad + 3\delta \cdot Tr_{k'/k}(\alpha) \cdot XY^2 + \delta \cdot Tr_{k'/k}(\alpha\beta) \cdot Y^3.
\end{aligned}
$$

Proof.   Let $k = \mathbb{F}_q$.

(1) It is

$$
\begin{aligned}
(\delta,\delta) &= \delta \cdot (1,1), \\
(\delta^2,\delta^2) &= \delta^2 \cdot (1,1), \\
(1,\delta^2) &= \tau[\delta^2 \cdot (1,\delta) \cdot (1,\tfrac{1}{\delta})^3] \text{ and} \\
(\delta,\delta^2) &= \delta \cdot (1,\delta),
\end{aligned}
$$

so corollary 9.2 implies that the classes $(L_1, (\delta, \delta))$ and $(L_1, (\delta^2, \delta^2))$ are equivalent to $(L_1, (1, 1))$ and that the classes $(L_1, (1, \delta^2))$ and $(L_1, (\delta, \delta^2))$ are equivalent to $(L_1, (1, \delta))$. Furthermore, it is easy to see that the classes $(L_1, (1, 1))$ and $(L_1, (1, \delta))$ are not equivalent.

To see that the classes $(L_\delta, 1)$, (L_δ, α) and (L_δ, α^2) are equivalent, we write $q = 3s + 1$ with $s \in \mathbb{N}_+$ and get

$$\overbrace{(\alpha^{s+1})^3}^{\in L_\delta^{\times 3}} \cdot 1 \cdot \frac{1}{\delta} = \alpha^{(3s+3)-([3s+1]+1)} = \alpha,$$

$$\underbrace{\left(\frac{1}{\alpha^s}\right)^3}_{\in L_\delta^{\times 3}} \cdot 1 \cdot \delta = \alpha^{-3s+([3s+1]+1)} = \alpha^2.$$

Now the claim follows from corollary 8.2.

(2) All we have to do according to corollary 8.2 is to show that $(L_\delta, 1)$ and (L_δ, α) are not equivalent: Write $q = 3s + 2$ with $s \in \mathbb{N}_+$. If $(L_\delta, 1)$ and (L_δ, α) were equivalent, then according to corollary 9.2 there would be numbers $u, v \in \mathbb{N}_0$ with $\alpha = (\alpha^u)^3 \cdot 1 \cdot \delta^v$ or $\alpha^q = (\alpha^u)^3 \cdot 1 \cdot \delta^v$. But such numbers can not exist, because

$$(\alpha^u)^3 \cdot 1 \cdot \delta^v = \alpha^{3u+(3s+2+1)v} = (\alpha^{u+(s+1)v})^3$$

is a third power whereas α and α^q are not. $\qquad\qquad\square$

Example 9.1 We want to continue example 7.2 and example 7.4 by computing the $\widetilde{\mathcal{F}_{\mathbb{C}}^{3,4}}/\widetilde{\mathcal{F}_{\mathbb{R}}^{3,4}}$-forms of P_3^4 as well as the $\widetilde{\mathcal{F}_{\mathbb{F}_5}^{4,3}}/\widetilde{\mathcal{F}_{\mathbb{F}_5}^{4,3}}$-forms of P_3^4:

(1) Proposition 9.2 implies that $(\mathbb{R}^3, (1, 1, 1))$ and $(\mathbb{R}^3, (-1, -1, -1))$ are equivalent, that $(\mathbb{R}^3, (1, 1, -1))$ and $(\mathbb{R}^3, (1, -1, -1))$ are equivalent and that $(\mathbb{C}^2 \times \mathbb{R}, (1, 1))$ and $(\mathbb{C}^2 \times \mathbb{R}, (1, -1))$ are equivalent. Consequently, there are exactly three $\widetilde{\mathcal{F}_{\mathbb{C}}^{3,4}}/\widetilde{\mathcal{F}_{\mathbb{R}}^{3,4}}$-forms of P_3^4, namely

$$E(\widetilde{\mathcal{F}_{\mathbb{C}}^{3,4}}/\widetilde{\mathcal{F}_{\mathbb{R}}^{3,4}})$$
$$= \{x^4 + y^4 + z^4,\ x^4 + y^4 - z^4,\ 2x^4 - 12x^2y^2 + 2y^4 + z^4\}.$$

(2) As mentioned earlier, taking the third power is *bijective* in $\mathbb{F}_5$, so that for every fcs-algebra L of degree four over $\mathbb{F}_5$ we have $k^\times \subseteq (L^\times)^3$.

This obviously implies

$$\mathrm{Aut}_{\mathbb{F}_5}(L)\backslash(L^\times/L^{\times^3})/\mathbb{F}_5^\times = \mathrm{Aut}_{\mathbb{F}_5}(L)\backslash(L^\times/L^{\times^3}),$$

and we see that all nine forms listed in example 7.4 are distinct in $\widetilde{\mathcal{F}_{\mathbb{F}_5}^{4,3}}$ as well.

Chapter 10

Representations of semidirect products

After having developed the method of Galois descent in the last chapters, we are now turning towards another "ingredient" we are going to need to compute the cohomology of twisted Fermat hypersurfaces:

The wreath product $\mu_r \wr S_n$ acts on the Fermat hypersurface X_n^r, and in the introduction we already mentioned that it is possible to use this fact to decompose the l-adic cohomology of X_n^r into certain eigenspaces corresponding to the characters of μ_r^n.

In this chapter, we will study the more general situation of a semidirect product $A \rtimes S$ acting on an object M of a (pseudo-)abelian category, and we will show that also in this general situation we will get a decomposition of M into eigenspaces M_χ corresponding to the characters χ of A.

In particular, this implies that the decomposition of $H_{\text{ét}}^*(\bar{X}_n^r, \mathbb{Q}_l)$ into eigenspaces is a "motivic" decomposition, the l-adic realization of the corresponding decomposition of the Grothendieck motif $h(X_n^r)$ of X_n^r.

Definition 10.1   Let A be a finite abelian group. A *character of A* is a homomorphism $\chi : A \to \mathbb{C}^\times$. The set of all characters is denoted by $\check{A}$, and it is becomes an abelian group, the so-called *dual of A*, by setting $(\chi\psi)(a) := \chi(a) \cdot \psi(a)$ for $\chi, \psi \in \check{A}$ and $a \in A$. — The unit element of $\check{A}$ is the *unit character* $1_{\check{A}}$ with $1_{\check{A}}(a) = 1$ for all $a \in A$.

Example 10.1   Let $r \in \mathbb{N}_+$ be a natural number. Then the dual of μ_r, the group of r-th roots of unity in $\mathbb{C}$, is canonically isomorphic to $\mathbb{Z}/r\mathbb{Z}$:

The isomorphism $\mathbb{Z}/r\mathbb{Z} \to \check{\mu}_r$ is given by sending $\bar{a}$ to $[\zeta \mapsto \zeta^a]$, and if $\chi \in \check{\mu}_r$ is an arbitrary character, then $\chi\left(e^{\frac{2\pi i}{r}}\right) = e^{\frac{2\pi i a}{r}}$ for $a \in \mathbb{Z}$, and sending χ to $\bar{a}$ gives the inverse $\check{\mu}_r \to \mathbb{Z}/r\mathbb{Z}$.

Remark 10.1　*It is a well known fact (compare for example [Ser73, VI.1]) that A is (noncanonically) isomorphic to $\check{A}$ and canonically isomorphic to $\check{\check{A}}$.*

Definition 10.2　Let $\varphi : A \to B$ be a homomorphism of finite abelian groups, then we define the *dual* homomorphism $\check{\varphi} : \check{B} \to \check{A}$ by setting

$$\check{\varphi}(\chi) := \left[A \xrightarrow{\varphi} B \xrightarrow{\chi} \mathbb{C}^\times \right]$$

for $\chi \in \check{B}$.

Proposition 10.1　*Sending a finite abelian group A to its dual $\check{A}$ and a homomorphism $\varphi : A \to B$ to its dual $\check{\varphi}$ defines a contravariant additive exact functor from the category of finite abelian groups into itself.*

Proof.　Denote the functor in question by $D : \underline{\mathrm{Ab}}^{\mathrm{fin}} \to \left(\underline{\mathrm{Ab}}^{\mathrm{fin}} \right)^{\mathrm{opp}}$, and denote the canonical embedding $\underline{\mathrm{Ab}}^{\mathrm{fin}} \to \underline{\mathrm{Ab}}$ by F. Then it is enough to prove that the composition $FD : \underline{\mathrm{Ab}}^{\mathrm{fin}} \to \underline{\mathrm{Ab}}^{\mathrm{opp}}$ is an additive exact functor. But $FD = \mathrm{Hom}_{\underline{\mathrm{Ab}}}(\_, \mathbb{C}^\times)$, and $\mathbb{C}^\times$ is divisible an hence an injective object of $\underline{\mathrm{Ab}}$, so the proposition follows.　□

Lemma 10.1　*Let S be a group, and let A be a finite abelian S-group. Then the dual abelian group $\check{A}$ is made into an S-group by setting*

$$\forall s \in S : \forall \chi \in \check{A} : \forall a \in A : \ [^s\chi](a) := \chi(^{s^{-1}}a). \tag{10.1}$$

Proof.　First we want to show that $^s\chi$ is really an element of $\check{A}$ for arbitrary $s \in S$ and $\chi \in \check{A}$. For this, we just have to check that $[^s\chi](ab) = [^s\chi](a) \cdot [^s\chi](b)$ for all $a, b \in A$:

$$[^s\chi](ab) = \chi(^{s^{-1}}(ab)) = \chi(^{s^{-1}}a\,^{s^{-1}}b) = \chi(^{s^{-1}}a) \cdot \chi(^{s^{-1}}b) = [^s\chi](a) \cdot [^s\chi](b).$$

We obviously have $^1\chi = \chi$ for all $\chi \in \check{A}$, and for arbitrary $s, t \in S$, $\chi \in \check{A}$ and $a \in A$ we have

$$^{st}\chi(a) = \chi(^{t^{-1}s^{-1}}a) = {}^t\chi(^{s^{-1}}a) = {}^s({}^t\chi)(a),$$

so that $\check{A}$ is indeed an S-set. To prove that it is an S-group, let $s \in S$, $\chi, \psi \in \check{A}$ and $a \in A$ be arbitrary elements. Then we get

$$\left[{}^s(\chi\psi)\right](a) = (\chi\psi)(^{s^{-1}}a) = \chi(^{s^{-1}}a) \cdot \psi(^{s^{-1}}a) = [{}^s\chi](a) \cdot [{}^s\psi](a) = \left[{}^s\chi\,{}^s\psi\right](a),$$

so $^s(\chi\psi) = {}^s\chi \cdot {}^s\psi$. This completes the proof of the lemma. $\qquad\square$

Lemma 10.2   *Let A be a finite abelian group of order n, and let $\chi \in \check{A}$ be a character. Then*

$$\sum_{a \in A} \chi(a) = \begin{cases} n \text{ if } \chi = 1_{\check{A}} \\ 0 \text{ otherwise} \end{cases}$$

Proof.   We cite the easy proof from [Ser73, VI.1.2]: The formula is obvious if χ is the unit character. If χ is not the unit character, choose $b \in A$ such that $\chi(b) \neq 1$. One has:

$$\chi(b) \cdot \sum_{a \in A} \chi(a) = \sum_{a \in A} \chi(ab) = \sum_{a \in A} \chi(a),$$

hence:

$$(\chi(b) - 1) \cdot \sum_{a \in A} \chi(a) = 0.$$

Since $\chi(b) \neq 1$, this implies $\sum_{a \in A} \chi(a) = 0$. $\qquad\square$

Definition 10.3   Let R be a (not necessarily commutative) ring with unit. A system $\{e_1, \ldots, e_s\}$ of elements of R is called a *complete system of idempotents of R* if

$$\forall i, j \in \{1, \ldots, s\} : \; e_i \cdot e_j = \left\{ \begin{array}{l} e_i \text{ if } i = j \\ 0 \text{ otherwise} \end{array} \right\}, \quad \text{and } e_1 + \ldots + e_s = 1.$$

Lemma 10.3   *Let A be a finite abelian group of order n. For every character $\chi \in \check{A}$ define*

$$p_\chi := \frac{1}{n} \sum_{a \in A} \chi(a)^{-1} a \in \mathbb{C}[A].$$

Then $(p_\chi)_{\chi \in \check{A}}$ is a complete system of idempotents in $\mathbb{C}[A]$.

If A has exponent $r \in \mathbb{N}_+$ (which means that $a^r = 1_A$ for all $a \in A$), and if $R \subseteq \mathbb{C}$ is a ring which contains the r-th roots of unity and in which n is invertible, then $\{p_\chi | \chi \in \check{A}\}$ is contained in $R[A]$ and is a complete system of idempotents in $R[A]$.

Proof.   This follows from general results of the representation theory of finite groups (compare for example [Lan93, XVIII.4]), but we want to give an easy direct proof: Let χ and ψ be arbitrary characters of A. We have

$$p_\chi \cdot p_\psi = \frac{1}{n^2} \cdot \sum_{a \in A} \underbrace{\left[\sum_{\substack{b,c \in A \\ bc=a}} \chi(b)^{-1} \cdot \psi(c)^{-1} \right]}_{=:c_a} a. \qquad (10.2)$$

Let us first calculate c_a for every $a \in A$:

$$c_a = \sum_{\substack{b,c \in A \\ bc=a}} \chi(b)^{-1} \cdot \psi(c)^{-1} = \sum_{b \in A} \chi(b)^{-1} \cdot \psi(ab^{-1})^{-1}$$

$$= \psi(a)^{-1} \cdot \sum_{b \in A} \chi(b)^{-1} \cdot \psi(b) = \psi(a)^{-1} \cdot \sum_{b \in A} [\psi \chi^{-1}](b)$$

$$\overset{\text{lemma10.2}}{=} \begin{cases} n \cdot \psi(a)^{-1} & \text{if } \chi = \psi \\ 0 & \text{otherwise.} \end{cases}$$

Substituting this into (10.2), we immediately get $p_\chi \cdot p_\psi = 0$ if $\chi \neq \psi$. In case $\chi = \psi$ we get

$$p_\psi^2 = \frac{1}{n^2} \sum_{a \in A} c_a \cdot a = \frac{n}{n^2} \sum_{a \in A} \psi(a)^{-1} \cdot a = p_\psi.$$

Now let A have exponent r, and let R be a ring with $\mathbb{Z}[\frac{1}{n}, \mu_r] \subseteq R \subseteq \mathbb{C}$. If $\chi \in \check{A}$ is an arbitrary character, then

$$\forall a \in A : [\chi(a)]^r = \chi(a^r) = \chi(1_A) = 1,$$

which shows that the image of χ is contained in $\mu_r \subseteq R$. Then it is obvious from the definition of p_χ that $p_\chi \in R[A]$, and since $\{p_\chi\}$ is a complete system of idempotents in $\mathbb{C}[A]$, it must obviously also be a complete system of idempotents in $R[A]$. $\qquad\square$

Lemma 10.4   *Let S be a group, let A be a finite abelian S-group of order n and exponent r, let $A \rtimes S$ be the semidirect product, and let R be a ring with $\mathbb{Z}[\frac{1}{n}, \mu_r] \subseteq R \subseteq \mathbb{C}$. For a character $\chi \in \check{A}$, we consider the idempotent p_χ defined in lemma 10.3 as an element of $R[A \rtimes S]$ via $R[A] \hookrightarrow R[A \rtimes S]$. Then the following equations hold in $R[A \rtimes S]$ for all $s \in S$, $a \in A$ and $\chi \in \check{A}$:*

$$ s \cdot p_\chi = p_{{}^s\chi} \cdot s \quad and \quad a \cdot p_\chi = \chi(a) \cdot p_\chi. $$

Proof.

$$ s \cdot p_\chi = s \cdot \frac{1}{n} \sum_{b \in A} \chi(b)^{-1} b = \frac{1}{n} \sum_{b \in A} \chi(b)^{-1} sb = \frac{1}{n} \sum_{b \in A} \chi(b)^{-1} [{}^s b] \cdot s $$

$$ = \frac{1}{n} \sum_{b \in A} \chi({}^{s^{-1}}[{}^s b])^{-1} [{}^s b] \cdot s = \frac{1}{n} \sum_{b \in A} {}^s\chi([{}^s b])^{-1} [{}^s b] \cdot s $$

$$ = \frac{1}{n} \sum_{b \in A} {}^s\chi(b)^{-1} b \cdot s = p_{{}^s\chi} \cdot s. $$

$$ a \cdot p_\chi = \frac{1}{n} \sum_{b \in A} \chi(b)^{-1} [ab] = \frac{1}{n} \sum_{b \in A} \chi(a^{-1}[ab])^{-1} [ab] $$

$$ = \frac{1}{n} \sum_{b \in A} \chi(a) \cdot \chi([ab])^{-1} [ab] $$

$$ = \frac{1}{n} \sum_{b \in A} \chi(a) \cdot \chi(b)^{-1} b = \chi(a) \cdot p_\chi. $$

$\qquad\square$

Definition 10.4   Let A be a finite abelian group of order n and exponent r, let R be a ring with $\mathbb{Z}[\mu_r, \frac{1}{n}] \subseteq R \subseteq \mathbb{C}$, let $\mathcal{M}$ be a pseudo-abelian

R-linear category[1], and let $M \in \mathrm{Ob}(\mathcal{M})$ be an object on which A acts (note that because A is abelian, we do not have to distinguish between left- and right-A-actions). Then the group homomorphism $A \to \mathrm{Aut}_{\mathcal{M}}(M)$ induces a morphism $R[A] \to \mathrm{End}_{\mathcal{M}}(M)$ of (not necessarily commutative) R-algebras, so that for $\chi \in \check{A}$, the image of p_χ in $\mathrm{End}_{\mathcal{M}}(M)$ is a projector. We denote this projector by p_χ^M or simply by p_χ again, and we write M_χ for its image (which exists according to corollary A.1).

Lemma 10.5 *In the situation of definition 10.4, consider the special case where $\mathcal{M}$ is the (abelian) category of T-modules for an R-algebra T. Then for each character $\chi \in \check{A}$, we have*

$$M_\chi = \left\{ x \in M \,\middle|\, \forall a \in A : a(x) = \chi(a) \cdot x \right\} \tag{10.3}$$

Proof. The inclusion "$\subseteq$" immediately follows from the right diagram in (10.5). to see the other inclusion, let x be an arbitrary element of the righthand side of (10.3). Then we have

$$p_\chi(x) = \left[\frac{1}{n} \sum_{a \in A} \chi(a^{-1}) \cdot a \right](x) = \frac{1}{n} \sum_{a \in A} \chi(a^{-1}) \cdot a(x)$$

$$= \frac{1}{n} \sum_{a \in A} \chi(a^{-1}) \cdot \chi(a) \cdot x = \frac{1}{n} \sum_{a \in A} x = \frac{n}{n} \cdot x = x.$$

so $x = p_\chi(x)$ lies in the image of p_χ which by definition is M_χ. $\qquad \square$

Corollary 10.1 *In the situation of lemma 10.4, let $\mathcal{M}$ be a pseudo-abelian R-linear category, and let $M \in \mathrm{Ob}(\mathcal{M})$ be an object which is endowed with a* right-$(A \rtimes S)$-*action. Then the canonical morphism*

$$\bigoplus_{\chi \in \check{A}} M_\chi \longrightarrow M, \tag{10.4}$$

is an isomorphism, and for arbitrary elements $s \in S$, $a \in A$ and $\chi \in \check{A}$, the following two diagrams commute in M and hence describe the automor-

---

[1]The definition and basic properties of pseudo-abelian categories can be found in the appendix.

phisms s and a with respect to the decomposition (10.4) *of M:*

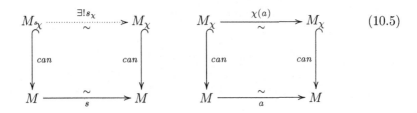

$$(10.5)$$

Proof. Since $\{p_\chi\}_{\chi \in \check{A}}$ is a complete system of idempotents in $R[A]$ by lemma 10.3, the same is true for its image in $\mathrm{End}_{\mathcal{M}}(M)$. So the fact that the morphism from (10.4) is an isomorphism follows from proposition A.2.

For $\chi \in \check{A}$, denote the canonical monomorphism $M_\chi \hookrightarrow M$ by α_χ, and denote the canonical epimorphism $M \twoheadrightarrow M_\chi$ by β_χ. So we have $\alpha_\chi \circ \beta_\chi = p_\chi$ and $\beta_\chi \circ \alpha_\chi = 1_{M_\chi}$ according to proposition A.1(2), and for $a \in A$ we get

$$[a \circ \alpha_\chi] \circ \beta_\chi = a \circ p_\chi \overset{\text{lemma 10.4}}{=} \chi(a) \circ p_\chi = [\chi(a) \circ \alpha_\chi] \circ \beta_\chi = [\alpha_\chi \circ \chi(a)] \circ \beta_\chi.$$

Since β_χ is an epimorphism, it follows that $a \circ \alpha_\chi = \alpha_\chi \circ \chi(a)$, so the right diagram in (10.5) commutes.

Let $\chi \in \check{A}$ and $s \in S$ be arbitrary elements. The composition of group homomorphisms

$$A \rtimes S \xrightarrow{x \mapsto x^{-1}} (A \rtimes S)^{\mathrm{op}} \to \mathrm{Aut}_{\mathcal{M}}(M)$$

induces a morphism of R-algebras

$$\varphi : R[A \rtimes S] \longrightarrow \mathrm{End}_{\mathcal{M}}(M),$$

and by definition we have $\varphi(p_\chi) = p_{\chi^{-1}}$ and $\varphi(s) = s^{-1}$. Then we calculate:

$$s \circ p_{s\chi} = \varphi\left(s^{-1} \cdot p_{(s\chi)^{-1}}\right) \overset{\text{lemma 10.4}}{=} \varphi\left(p_{(s^{-1})((s\chi)^{-1})} \cdot s^{-1}\right)$$
$$= \varphi\left(p_{\chi^{-1}} \cdot s^{-1}\right) = p_\chi \circ s.$$

With this we get:

$$\left[s \circ \alpha_{*_\chi}\right] \circ \beta_{*_\chi} = s \circ p_{*_\chi} = p_\chi \circ s = p_\chi \circ p_\chi \circ s$$

$$= p_\chi \circ s \circ p_{*_\chi} = \left[\alpha_\chi \circ \underbrace{\left(\beta_\chi \circ s \circ \alpha_{*_\chi}\right)}_{=:s_\chi}\right] \circ \beta_{*_\chi},$$

and because β_{*_χ} is an epimorphism, we see that the left diagram in (10.5) also commutes. $\qquad\Box$

Example 10.2 Let S be the symmetric group S_2, let

$$A := \mu_2^2 = \Big\{(1,1),(1,-1),(-1,1),(-1,-1)\Big\} \subset \mathbb{C}^\times \times \mathbb{C}^\times,$$

considered as an S_2-group in the obvious way, let $R := \mathbb{Q}$ (note that A has exponent two and that $\mathbb{Z}[\mu_2, \frac{1}{4}] \subseteq \mathbb{Q} \subseteq \mathbb{C}$), let $\mathcal{M}$ be the category of $\mathbb{Q}$-vector spaces, and let M be the vector space $\mathbb{Q}^2$, its elements considered as *row* vectors.

Then $A \rtimes S = \mu_2 \wr S_2$ is a subgroup of $\mathrm{GL}(2, \mathbb{Q})$ via lemma 7.1, and $\mathrm{GL}(2, \mathbb{Q})$ acts on $\mathbb{Q}^2$ by multiplication on the right.

By example 10.1 and proposition 10.1, the dual of A is canonically isomorphic to

$$(\mathbb{Z}/2)^2 = \Big\{[0,0],[0,1],[1,0],[1,1]\Big\}$$

with

$$\check{A} \quad \times \quad A \quad \longrightarrow \quad \mathbb{C}^\times$$
$$\big([\bar{a},\bar{b}] \ , \ (\zeta_1,\zeta_2)\big) \quad \mapsto \quad \zeta_1^a \cdot \zeta_2^b.$$

We thus get the following table listing the results of the pairing $\check{A} \times A \to \mathbb{C}^\times$:

	(1,1)	(1,-1)	(-1,1)	(-1,-1)
$[0,0]$	1	1	1	1
$[0,1]$	1	-1	1	-1
$[1,0]$	1	1	-1	-1
$[1,1]$	1	-1	-1	1

With this table, we can compute the projectors p_χ:

$$p_{[0,0]} = \tfrac{1}{4} \left[1 \cdot \left(\begin{smallmatrix} 1 & 0 \\ 0 & 1 \end{smallmatrix} \right) + 1 \cdot \left(\begin{smallmatrix} 1 & 0 \\ 0 & -1 \end{smallmatrix} \right) + 1 \cdot \left(\begin{smallmatrix} -1 & 0 \\ 0 & 1 \end{smallmatrix} \right) + 1 \cdot \left(\begin{smallmatrix} -1 & 0 \\ 0 & -1 \end{smallmatrix} \right) \right] = \quad 0,$$

$$p_{[0,1]} = \tfrac{1}{4} \left[1 \cdot \left(\begin{smallmatrix} 1 & 0 \\ 0 & 1 \end{smallmatrix} \right) - 1 \cdot \left(\begin{smallmatrix} 1 & 0 \\ 0 & -1 \end{smallmatrix} \right) + 1 \cdot \left(\begin{smallmatrix} -1 & 0 \\ 0 & 1 \end{smallmatrix} \right) - 1 \cdot \left(\begin{smallmatrix} -1 & 0 \\ 0 & -1 \end{smallmatrix} \right) \right] = \left(\begin{smallmatrix} 0 & 0 \\ 0 & 1 \end{smallmatrix} \right),$$

$$p_{[0,1]} = \tfrac{1}{4} \left[1 \cdot \left(\begin{smallmatrix} 1 & 0 \\ 0 & 1 \end{smallmatrix} \right) + 1 \cdot \left(\begin{smallmatrix} 1 & 0 \\ 0 & -1 \end{smallmatrix} \right) - 1 \cdot \left(\begin{smallmatrix} -1 & 0 \\ 0 & 1 \end{smallmatrix} \right) - 1 \cdot \left(\begin{smallmatrix} -1 & 0 \\ 0 & -1 \end{smallmatrix} \right) \right] = \left(\begin{smallmatrix} 1 & 0 \\ 0 & 0 \end{smallmatrix} \right),$$

$$p_{[1,1]} = \tfrac{1}{4} \left[1 \cdot \left(\begin{smallmatrix} 1 & 0 \\ 0 & 1 \end{smallmatrix} \right) - 1 \cdot \left(\begin{smallmatrix} 1 & 0 \\ 0 & -1 \end{smallmatrix} \right) - 1 \cdot \left(\begin{smallmatrix} -1 & 0 \\ 0 & 1 \end{smallmatrix} \right) + 1 \cdot \left(\begin{smallmatrix} -1 & 0 \\ 0 & -1 \end{smallmatrix} \right) \right] = \quad 0.$$

So we have $M_{[0,0]} = 0$, $M_{[0,1]} = \langle (0,1) \rangle$, $M_{[1,0]} = \langle (1,0) \rangle$ and $M_{[1,1]} = 0$.

Now let $\chi := [0,1]$, let $a := (1,-1)$, and let s be the transposition (12). Then

$$(0,1) \cdot a = (0,1) \cdot \left(\begin{smallmatrix} 1 & 0 \\ 0 & -1 \end{smallmatrix} \right) = (0,-1),$$

so on M_χ, the automorphism a is multiplication by $\chi(a) = -1$ as it should be according to the right diagram in 10.5. Furthermore, we have ${}^s\chi = [1,0]$ and

$$(1,0) \cdot s = (1,0) \cdot \left(\begin{smallmatrix} 0 & 1 \\ 1 & 0 \end{smallmatrix} \right) = (0,1),$$

so in accordance with the left diagram in 10.5, the subspace $M_{{}^s\chi}$ is mapped to M_χ under s.

Lemma 10.6 *In the situation of corollary 10.1, let $\chi \in \check{A}$ be an arbitrary character, and denote the orbit of χ under the action of S defined in lemma 10.1 by $[\chi] \subseteq \check{A}$. Put*

$$M_{[\chi]} := \bigoplus_{\psi \in [\chi]} M_\psi.$$

If $\check{A} = [\chi_1] \sqcup \ldots \sqcup [\chi_m]$ is a decomposition into disjoint S-orbits, then

$$M \cong M_{[\chi_1]} \oplus \ldots \oplus M_{[\chi_m]}$$

is a decomposition of M into $(A \rtimes S)$-invariant subobjects.

Proof. We only have to show that for $\chi \in \check{A}$, the subobject $M_{[\chi]}$ is $(A \rtimes S)$-invariant. This means that for all $a \in A$, $s \in S$ and $\psi \in [\chi]$, the morphism

$$M_\psi \overset{\text{can}}{\hookrightarrow} M \overset{a \cdot s}{\longrightarrow} M \tag{10.6}$$

factors over $M_{[\chi]}$. By definition of $[\chi]$, we have $\psi = {}^t\chi$ for a suitable $t \in S$, and by corollary 10.1, we have a commutative diagram

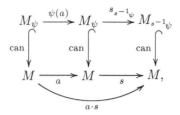

so the morphism from (10.6) factors over $M_{s^{-1}\psi}$. But

$$s^{-1}\psi = {}^{s^{-1}}({}^t\chi) = {}^{(s^{-1}t)}\chi \in [\chi],$$

and we are done.   □

Example 10.3   In example 10.2, we have

$$[[0,0]] = \Big\{[0,0]\Big\},$$
$$[[0,1]] = \Big\{[0,1],[1,0]\Big\},$$
$$[[1,1]] = \Big\{[1,1]\Big\},$$

so $M_{[[0,0]]} = M_{[[1,1]]} = 0$ and $M_{[[0,1]]} = M_{[0,1]} \oplus M_{[1,0]} = M$.

Example 10.4   Let S be a group, let A be a finite abelian S-group of exponent r, let k be a field with absolute Galois group G_k, and let X be a smooth projective variety over k, endowed with a *left-*$(A \rtimes S)$-action.

(1) Let E be a field with $\mathbb{Z}[\mu_r, \frac{1}{\#A}] \subset E \subseteq \mathbb{C}$, and let $\mathcal{M}$ be the category $\mathcal{M}_k^E$ of Grothendieck motives over k with coefficients in E.
    Then by functorality, the motif $h(X) \in \mathrm{Ob}(\mathcal{M})$ of X is endowed with a right-$(A \rtimes S)$-action, and by corollary 10.1 we get a canonical decomposition

$$h(X) \cong \bigoplus_{\chi \in \check{A}} h(X)_\chi.$$

(2) Let $l \neq \mathrm{char}(k)$ be a prime number with $l \equiv 1 \pmod{r}$, and let $i \in \mathbb{N}_0$ be a natural number.

By Hensel's lemma, $\mathbb{Q}_l$ contains the r-th roots of unity, so after choosing a primitive r-th root of unity in $\mathbb{Q}_l$, we get an embedding $\mathbb{Q}(\zeta) \hookrightarrow \mathbb{Q}_l$ which gives $\mathcal{M} := \mathbf{Rep}_{\mathbb{Q}_l}^{G_k}$ the structure of a $\mathbb{Q}(\zeta)$-linear abelian category.

Denote the $\mathbb{Q}_l$-G_k-representation $\mathrm{H}^i_{\text{ét}}(X \times_k \bar{k}, \mathbb{Q}_l)$ by $V \in \mathrm{Ob}(\mathcal{M})$. Then by functorality, V is endowed with a right-$(A \rtimes S)$-action, and corollary 10.1 implies that we have a direct sum decomposition $V \cong \bigoplus V_\chi$ of V into $\mathbb{Q}_l$-G_k-representations V_χ, where according to lemma 10.5 we have

$$V_\chi = \left\{ v \in V \,\middle|\, \forall a \in A : a(v) = \chi(a) \cdot v \right\}$$

for each character $\chi \in \check{A}$.

Lemma 10.7 *In the situation of definition 10.4, let N be a second object of M on which A acts, let $f \in \mathrm{Mor}_{\mathcal{M}}(M, N)$ be an A-equivariant morphism, and let $\chi \in \check{A}$ be an arbitrary character. Then we get the following induced diagram of A-equivariant morphisms in $\mathcal{M}$:*

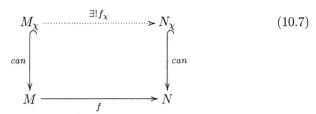

$$(10.7)$$

If furthermore S is an arbitrary group acting on A, and if M and N are endowed with right-$(A \rtimes S)$-actions such that f is $(A \rtimes S)$-equivariant, then we get the following induced diagram of $(A \rtimes S)$-equivariant morphisms in $\mathcal{M}$:

$$
\begin{array}{ccc}
M_{[\chi]} & \xrightarrow{\;\exists! f_{[\chi]}\;} & N_{[\chi]} \\
\Big\uparrow{\scriptstyle can} & & \Big\uparrow{\scriptstyle can} \\
M & \xrightarrow[\;f\;]{} & N.
\end{array}
\qquad (10.8)
$$

Proof. The morphisms $\alpha_\chi : M_\chi \hookrightarrow M$ and $\alpha_\chi : N_\chi \hookrightarrow N$ are A-equivariant because of corollary 10.1, and the morphisms $\alpha_{[\chi]} : M_{[\chi]} \hookrightarrow M$ and $\alpha_{[\chi]} : N_{[\chi]} \hookrightarrow N$ are $(A \rtimes S)$-equivariant because of lemma 10.6. In addition to that, for each $\psi \in \check{A}$ we have

$$
f \circ p_\psi = f \circ \left(\frac{1}{n} \sum_{a \in A} \psi(a)^{-1} a \right) = \frac{1}{n} \sum_{a \in A} \psi(a)^{-1} (f \circ a) =
$$
$$
\frac{1}{n} \sum_{a \in A} \psi(a)^{-1}(a \circ f) = \left(\frac{1}{n} \sum_{a \in A} \psi(a)^{-1} a \right) \circ f = p_\psi \circ f. \quad (10.9)
$$

If we denote the epimorphisms $M \twoheadrightarrow M_\chi$ and $N \twoheadrightarrow N_\chi$ by β_χ and the epimorphisms $M \twoheadrightarrow M_{[\chi]}$ and $N \twoheadrightarrow N_{[\chi]}$ by $\beta_{[\chi]}$, we therefore get

$$
[f \circ \alpha_\chi] \circ \beta_\chi = f \circ p_\chi \overset{(10.9)}{=} p_\chi \circ f = p_\chi \circ p_\chi \circ f
$$
$$
\overset{(10.9)}{=} p_\chi \circ [f \circ p_\chi] = \alpha_\chi \circ \underbrace{\beta_\chi \circ f \circ \alpha_\chi}_{=:f_\chi} \circ \beta_\chi = [\alpha_\chi \circ f_\chi] \circ \beta_\chi
$$

and hence $f \circ \alpha_\chi = \alpha_\chi \circ f$, because β_χ is an epimorphism. This proves that diagram (10.7) commutes. For an arbitrary element $a \in A$, we have

$$
f_\chi \circ a \overset{\text{corollary } 10.1}{=} f_\chi \circ \chi(a) = \chi(a) \circ f_\chi \overset{\text{corollary } 10.1}{=} a \circ f_\chi,
$$

so f_χ is A-equivariant, and the first part of the lemma is proven. For the second part, put $f_{[\chi]} := \bigoplus_{\psi \in [\chi]} f_\psi$. Then diagram (10.8) commutes, because diagram (10.7) commutes, and we only have to show that $f_{[\chi]}$ is $(A \rtimes S)$-equivariant. For that, let $a \cdot s$ be an arbitrary element of $A \rtimes S$.

Then because of corollary 10.1, the following two diagrams in $\mathcal{M}$ commute:

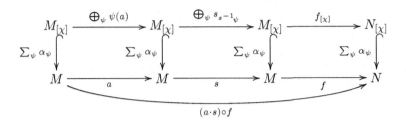

$$(a \cdot s) \circ f$$

so because $\sum_\psi \alpha_\psi$ is a monomorphism, we have

$$f \circ \left[\sum s_{s^{-1}\psi} \circ \sum \psi(a) \right] = \left[\sum s_{s^{-1}\psi} \circ \sum \psi(a) \right] \circ f$$

and hence

$$f \circ \left[(a \cdot s) | M_{[\chi]} \right] = \left[(a \cdot s) | N_{[\chi]} \right] \circ f.$$

$\square$

Lemma 10.8   *Let A be a finite abelian group of order n and exponent r, let R be a ring with $\mathbb{Z}[\mu_r, \frac{1}{n}] \subseteq R \subseteq \mathbb{C}$, let $\varphi : A \twoheadrightarrow B$ be an epimorphism of finite abelian groups, and let χ be an arbitrary character of B.*

Then under the morphism $R[A] \xrightarrow{\varphi_*} R[B]$ of R-algebras induced by φ, the idempotent $p_{\hat{\varphi}(\chi)}$ is mapped to p_χ.

Proof.   Put $C := \text{Ker}(\varphi)$, and let $\mathcal{B} \subseteq A$ be a system of representatives for B. Then we get:

$$\varphi_*[p_{\check{\varphi}(\chi)}] = \varphi_*\left(\frac{1}{n}\sum_{a\in A}\check{\varphi}(\chi)(a^{-1})a\right) = \varphi_*\left(\frac{1}{n}\sum_{a\in A}\chi(\varphi a)^{-1}a\right)$$

$$= \frac{1}{n}\sum_{a\in A}\chi(\varphi a)^{-1}\varphi a = \frac{1}{n}\sum_{c\in C}\sum_{b\in B}\chi(\varphi(cb))^{-1}\varphi(cb)$$

$$= \frac{\#C}{n}\sum_{b\in B}\chi(\varphi b)^{-1}\varphi b = \frac{1}{\#B}\sum_{b\in B}\chi(b)^{-1}b = p_\chi.$$

$\square$

Corollary 10.2 *Let A be a finite abelian group of order n and exponent r, let R be a ring satisfying $\mathbb{Z}[\mu_r, \frac{1}{n}] \subseteq R \subseteq \mathbb{C}$, and let N be an object of $\mathcal{M}$ equipped with an A-action.*

Furthermore, let $\varphi : A \twoheadrightarrow B$ be an epimorphism of finite abelian groups, let M be an object of $\mathcal{M}$ equipped with a B-action (which then also carries an A-action via $A \xrightarrow{\varphi} B \to \mathrm{Aut}_{\mathcal{M}}(M)$), and let $\chi \in \check{B}$ be an arbitrary character. Finally, let $f : M \to N$ be an A-equivariant morphism in $\mathcal{M}$. Then:

(1) We have a commuting diagram with A-equivariant morphisms as follows:

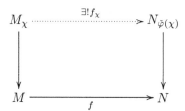

(2) Let S be a group, and assume that A is an S-group in such a way that $\mathrm{Ker}(\varphi)$ is S-invariant. Then there is a unique structure of an S-group on B that turns φ into a morphism of S-groups, and we get an induced group homomorphism $A \rtimes S \xrightarrow{\varphi_*} B \rtimes S$. Assume further that the given actions of A on N respectively of B on M are induced by a right-$(A \rtimes S)$-action on N respectively a right-$(B \rtimes S)$-action on M, and assume that f is even $(A \rtimes S)$-equivariant. Then we have a

commuting diagram with $(A \rtimes S)$-equivariant morphisms as follows:

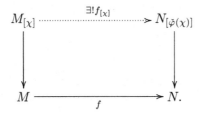

Proof. Applying lemma 10.7 to the character $\check{\varphi}(\chi)$, we see that for (1) we only have to show $p_\chi^M = p_{\check{\varphi}(\chi)}^M$. But this is easy, because $p_{\check{\varphi}(\chi)}^M$ is by definition the image of $p_{\check{\varphi}(\chi)}$ under

$$R[A] \longrightarrow R[B] \longrightarrow \mathrm{End}_{\mathcal{M}}(M),$$

which is the same as the image of p_χ under $R[B] \to \mathrm{End}_{\mathcal{M}}(M)$ by lemma 10.8.

Now let S be a group as in (2). The existence of a unique S-group-structure on B making φ a morphism of S-groups follows from lemma 4.6, and it is obvious that we get an induced group homomorphism $A \rtimes S \to B \rtimes S$. Then (2) follows exactly as (1) from lemma 10.7 and lemma 10.8. $\qquad\square$

Chapter 11

The l-adic cohomology of Fermat varieties

In this chapter, we are going to study the cohomology of twisted Fermat hypersurfaces over finite fields, and for any automorphism of X_n^r, we will compute the induced automorphism on cohomology. We will only be interested in the *middle* cohomology, because in the case of hypersurfaces, this is the only interesting one — the others are simply generated by the powers of the class of a hyperplane section.

Let $n, r \in \mathbb{N}_+$ with $n \geq 2$, let $k = \mathbb{F}_q$ be a finite field of characteristic $p > \max(r, 2)$ with separable closure K and absolute Galois group G_k, and let $P_n^r \in \mathrm{Ob}(\widetilde{\mathcal{F}_k^{n,r}})$ be the Fermat equation over k already studied in chapters 7 and 9.

As in example 5.1(2), we consider the $(n-2)$-dimensional hypersurface associated to P_n^r which we denote by $\mathcal{X} := \mathcal{X}_n^r$. So $\mathcal{X}$ is the hypersurface defined by the homogenous equation $X_1^r + \ldots + X_n^r = 0$ in $\mathbb{P}_k^{n-1}$.

Let $\bar{\mathcal{X}} := \bar{\mathcal{X}}_n^r := \mathcal{X} \times_k K$, and let F be the geometric Frobenius endomorphism on $\bar{\mathcal{X}}$. We want to study the action of the induced automorphism F^* on the middle geometric l-adic cohomology of $\mathcal{X}$ for a prime $l \neq p$ satisfying $l \equiv 1 \pmod{r}$, i.e. the $\mathbb{Q}_l$-G_k-representation $V := \mathrm{H}_{\text{ét}}^{n-2}(\bar{\mathcal{X}}, \mathbb{Q}_l)$.

Let μ_r be the group of r-th roots of unity in K, let $\tilde{A}$ be the group $(\mu_r^n/\mu_r) \rtimes S_n$ defined in proposition 9.1 respectively lemma 9.1, let $\zeta \in \mathbb{C}$ be the primitive r-th root of unity $e^{\frac{2\pi i}{r}}$, and fix an embedding $\mu_r \hookrightarrow \mathbb{Q}(\zeta)$.

Lemma 11.1 *Let A be the finite abelian group μ_r^n/μ_r (hence $\tilde{A} = A \rtimes S_n$).*

Then the dual $\check{A}$ of A is canonically isomorphic to

$$\left\{ \boldsymbol{a} = (a_1, \ldots, a_n) \in (\mathbb{Z}/r\mathbb{Z})^n \;\middle|\; \sum_{i=1}^n a_i = 0 \in \mathbb{Z}/r\mathbb{Z} \right\}. \tag{11.1}$$

With respect to this description, the pairing $\check{A} \times A \to \mathbb{C}^\times$ is given by

$$
\begin{array}{ccccc}
\check{A} & \times & A & \longrightarrow & \mathbb{C}^\times \\
\left((\bar{a}_1, \ldots, \bar{a}_n) \right. & , & \overline{(\zeta_1, \ldots, \zeta_n)} \left. \right) & \mapsto & \zeta_1^{a_1} \cdot \ldots \cdot \zeta_n^{a_n}
\end{array} \tag{11.2}
$$

and the S_n-action defined in lemma 10.1 by

$${}^s(a_1, \ldots, a_n) = (a_{s^{-1}(1)}, \ldots, a_{s^{-1}(n)}). \tag{11.3}$$

As in chapter 10, for $\boldsymbol{a} \in \check{A}$ we denote the orbit of $\boldsymbol{a}$ under the S_n-action by $[\boldsymbol{a}]$.

Proof. Because dualizing is additive and exact by proposition 10.1, the exact sequence

$$1 \to \mu_r \xrightarrow{\Delta} \mu_r^n \to A \to 1$$

induces an exact sequence

$$0 \to \check{A} \to (\check{\mu}_r)^n \xrightarrow{\check{\Delta}} \check{\mu}_r \to 0.$$

According to example 10.1, this sequence is canonically isomorphic to the sequence

$$0 \to \check{A} \to (\mathbb{Z}/r\mathbb{Z})^n \xrightarrow{\check{\Delta}} \mathbb{Z}/r\mathbb{Z} \to 0.$$

Since $\Delta = \prod_{i=1}^n 1_{\mu_r}$, we have $\check{\Delta} = \sum_{i=1}^n 1_{\mathbb{Z}/r\mathbb{Z}}$, and it follows that $\check{A}$ is indeed canonically isomorphic to the group defined in (11.1). Description (11.2) then follows immediately from the description of the isomorphism $\mathbb{Z}/r\mathbb{Z} \cong \check{\mu}_r$ given in example 10.1, and formula (11.3) easily follows from this:

$$
\begin{aligned}
\left[{}^s(a_1, \ldots, a_n)\right](\zeta_1, \ldots, \zeta_n) &\overset{(10.1)}{=} (a_1, \ldots, a_n)\left[{}^{s^{-1}}(\zeta_1, \ldots, \zeta_n)\right] \\
&= (a_1, \ldots, a_n)(\zeta_{s(1)}, \ldots, \zeta_{s(n)}) \overset{(11.2)}{=} \zeta_{s(1)}^{a_1} \cdot \ldots \cdot \zeta_{s(n)}^{a_n} \\
&= \zeta_1^{a_{s^{-1}(1)}} \cdot \ldots \cdot \zeta_n^{a_{s^{-1}(n)}} = (a_{s^{-1}(1)}, \ldots, a_{s^{-1}(n)})(\zeta_1, \ldots, \zeta_n).
\end{aligned} \tag{11.4}
$$

Definition 11.1 We define an action of $(\mathbb{Z}/r\mathbb{Z})^\times$ on $\check{A}$ by setting $t\boldsymbol{a} := (ta_1, \ldots, a_n)$ for $t \in (\mathbb{Z}/r\mathbb{Z})^\times$ and $\boldsymbol{a} \in \check{A}$, and for $\boldsymbol{a} \in \check{A}$, we denote the subset $\{q^i\boldsymbol{a} | i \in \mathbb{N}_0\}$ of $\check{A}$ by $\langle \boldsymbol{a} \rangle$ (so if k contains the r-th roots of unity, we have $\langle \boldsymbol{a} \rangle = \{\boldsymbol{a}\}$). Furthermore, we define the group

$$A_n^r := \left\{ \boldsymbol{a} \in \check{A} \mid \forall i \in \{1, \ldots, n\} \; : \; a_i \neq 0 \in \mathbb{Z}/r\mathbb{Z} \right\}.$$

Obviously the actions of S_n and $(\mathbb{Z}/r\mathbb{Z})^\times$ on $\check{A}$ leave A_n^r invariant and therefore restrict to actions on A_n^r.

Lemma 11.2 *According to example 10.4(2), we get a canonical decomposition of V in the category of $\mathbb{Q}_l$-vector spaces*

$$V = \bigoplus_{\boldsymbol{a} \in \check{A}} V_{\boldsymbol{a}}. \tag{11.5}$$

For $\boldsymbol{a} \in \check{A}$, we define (compare definition 10.6):

$$V_{[\boldsymbol{a}]} := \bigoplus_{\boldsymbol{b} \in [\boldsymbol{a}]} V_{\boldsymbol{b}}, \quad V_{\langle \boldsymbol{a} \rangle} := \bigoplus_{\boldsymbol{b} \in \langle \boldsymbol{a} \rangle} V_{\boldsymbol{b}}.$$

Proof. We can apply example 10.4(2) to the case $S := S_n$, $X := \mathcal{X}$ and $i := n - 2$: By proposition 9.1, the group $S \ltimes A = \tilde{A}$ acts on P_n^r from the left, so by functorality $\mathcal{X}$ is endowed with a left $(S \ltimes A)$-action, and we get decomposition (11.5). $\qquad\square$

Lemma 11.3 *Consider $\bar{\mathcal{X}}$ as a k-scheme, let f be the arithmetic Frobenius on $\bar{\mathcal{X}}$, and define a $\mathbb{Z}$-action on A by*

$$1 \mapsto \left[(\zeta_1, \ldots, \zeta_n) \mapsto \left(\zeta_1^{1/q}, \ldots, \zeta_n^{1/q} \right) \right] \in Aut(A)$$

(note that q is a unit modulo r). Then

$$A \rtimes \mathbb{Z} \ni a \cdot c \mapsto a \circ f^c \in Aut_k(\bar{\mathcal{X}})$$

defines a left-$(A \rtimes \mathbb{Z})$-action on $\bar{\mathcal{X}}$.

Proof.   Let $(\zeta_1, \ldots, \zeta_n)$ be an arbitrary element of A. Then the following diagram obviously commutes:

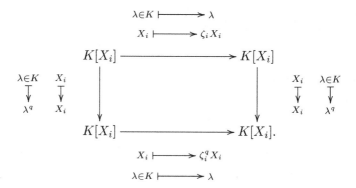

This implies that the following diagram commutes as well:

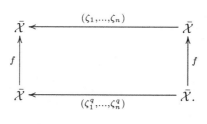

From this, by an easy induction, we get that

$$\forall (\zeta_1, \ldots, \zeta_n) \in A: \ \forall c \in \mathbb{Z}:$$
$$f^c \circ (\zeta_1, \ldots, \zeta_n) = \left(\zeta_1^{1/q^c}, \ldots, \zeta_n^{1/q^c} \right) \circ f^c \in \mathrm{Aut}_k(\bar{\mathcal{X}}). \quad (11.6)$$

Let $\varphi: A \rtimes \mathbb{Z} \to \mathrm{Aut}_k(\bar{\mathcal{X}})$ be the map defined by $\varphi(a \cdot c) := a \circ f^c$ as in the lemma. We claim that φ is a left-$(A \rtimes \mathbb{Z})$-action on $\bar{\mathcal{X}}$. To check this, let $a = (\zeta_i)$, $b = (\xi_i) \in A$ and $c, d \in \mathbb{Z}$ be arbitrary elements. Then

$$\varphi\Big[(a \cdot c) \cdot (b \cdot d) \Big] = \varphi\Big[(a\,{}^c\!b) \cdot (c + d) \Big] = \varphi\Big[\left(\zeta_i \xi_i^{1/q^c} \right) \cdot (c + d) \Big]$$
$$= \left(\zeta_i \xi_i^{1/q^c} \right) \circ f^{c+d} = (\zeta_i) \circ \Big[\left(\xi_i^{1/q^c} \right) \circ f^c \Big] \circ f^d$$
$$\overset{(11.6)}{=} (\zeta_i) \circ \Big[f^c \circ (\xi_i) \Big] \circ f^d = \Big[(\zeta_i) \circ f^c \Big] \circ \Big[(\xi_i) \circ f^d \Big]$$
$$= \varphi(a \cdot c) \circ \varphi(b \cdot d),$$

so φ really defines a left-$(A \rtimes \mathbb{Z})$-action on $\bar{\mathcal{X}}$. $\square$

Corollary 11.1 *Let a be an arbitrary element of $\breve{A}$. Then F^* maps V_a to V_{qa}.*

Proof. Let $\mathcal{M}$ be the category of $\mathbb{Q}_l$-vector spaces, and let M be the object V of $\mathcal{M}$. According to lemma 11.3, we get an induced right-$(A \rtimes \mathbb{Z})$-action on M (compare example 10.4(2)) with $1_A \cdot (-1) = F^*$ by construction. We claim that the $\mathbb{Z}$-action on $\breve{A}$ defined in equation (10.1) is given by $1 \mapsto [b \mapsto qb]$. To prove this, let $b \in \breve{A}$ and $(\zeta_i) \in A$ be arbitrary elements. Then

$$[^1 b]\,(\zeta_i) \stackrel{(10.1)}{=} b\left(^{(-1)}(\zeta_i)\right) = b(\zeta_i^q) \stackrel{(11.2)}{=} \zeta_1^{qb_1} \cdot \ldots \cdot \zeta_n^{qb_n} = [qb](\zeta_i),$$

so we really have $^1 b = qb$.

We can therefore apply corollary 10.1 to $\chi := qa$ and $s := 11_A \cdot (-1) = F^*$ and get that the following diagram commutes:

$$
\begin{array}{ccc}
M_{s\chi} = V_a & \xrightarrow{\;\; F^*_{qa} \;\;} & M_\chi = V_{qa} \\[2pt]
{\scriptstyle\text{can}} \Big\uparrow & & \Big\uparrow {\scriptstyle\text{can}} \\[2pt]
M = V & \xrightarrow[\;\; F^* \;\;]{\sim} & M = V.
\end{array}
$$

This proves the corollary. $\square$

Corollary 11.2 *For all $a \in \breve{A}$, the subspace $V_{\langle a \rangle}$ is invariant under the Frobenius action and is thus a $\mathbb{Q}_l$-G_k-representation, so that we get the following decomposition of V in $\mathbf{Rep}_{\mathbb{Q}_l}^{G_k}$:*

$$V = \bigoplus_{\langle a \rangle} V_{\langle a \rangle}. \tag{11.7}$$

In particular, if k contains the r-th roots of unity (and hence $q \equiv 1$ (mod r)), decomposition (11.5) is a decomposition in $\mathbf{Rep}_{\mathbb{Q}_l}^{G_k}$.

Proof. According to corollary 11.1, the geometric Frobenius maps V_a to $V_{qa} \subseteq V_{\langle a \rangle}$ for all $a \in \breve{A}$. Since the arithmetic Frobenius, whose effect on cohomology is $(F^*)^{-1}$, topologically generates G_k, this proves that

$\varphi^*(V_{\boldsymbol{a}}) \subseteq V_{\langle \boldsymbol{a} \rangle}$ for all $\varphi \in G_k$, so decomposition (11.7) is a decomposition of $\mathbb{Q}_l$-G_k-representations.

In the special case $q \equiv 1 \pmod{r}$, we have $\langle \boldsymbol{a} \rangle = \{\boldsymbol{a}\}$ for all $\boldsymbol{a} \in \check{A}$, so decompositions (11.7) and (11.5) coincide. $\qquad\square$

Before we can state Deligne's result on the Frobenius action on V, we first have to define *Jacobi sums*:

Definition 11.2 Let $\kappa \subseteq K$ be a finite field which contains the r-th roots of unity. Then $\#\kappa \equiv 1 \pmod{r}$, so we can define a surjective multiplicative character $\kappa^\times \twoheadrightarrow \mu_r$ by sending $x \in \kappa^\times$ to $x^{\frac{\#\kappa-1}{r}} \in \mu_r$. Composition with the fixed embedding $\mu_r \hookrightarrow \mathbb{Q}(\zeta)$ thus gives us a group homomorphism $\kappa^\times \xrightarrow{\chi} \mathbb{Q}(\zeta)^\times$.

For $\boldsymbol{a} \in A_n^r$, we can then define the *Jacobi sum of dimension $(n-2)$ and degree r for $\boldsymbol{a}$ over κ* (which depends on the chosen embedding $\mu_r \hookrightarrow \mathbb{Q}(\zeta) \hookrightarrow \mathbb{Q}_l$) as

$$\mathcal{J}_\kappa^r(\boldsymbol{a}) := \mathcal{J}(\boldsymbol{a}) := (-1)^n \sum_{\substack{(v_2,\ldots,v_n)\in(\kappa^\times)^{n-1} \\ 1+v_2+\ldots+v_n=0}} \left[\prod_{i=2}^{n} \chi(v_i)^{a_i}\right]. \qquad (11.8)$$

We can consider $\mathcal{J}(\boldsymbol{a})$ as an element of both $\mathbb{Q}(\zeta)$ and $\mathbb{Q}_l$.

Lemma 11.4 *If r is odd, and if k contains the r-th roots of unity, then for all $\boldsymbol{a} \in A_2^r$ we have*

$$\mathcal{J}_k^r(\boldsymbol{a}) = 1.$$

Proof. By definition, we have

$$\mathcal{J}_k^r(\boldsymbol{a}) = (-1)^2 \sum_{\substack{v_2\in k^\times \\ 1+v_2=0}} \left[\prod_{i=2}^{2} \chi(v_2)^{a_2}\right] = \chi(-1)^{a_2}.$$

But (-1) has order two in $k^\times$, so we have $\chi(-1) \in \{-1,1\} \cap \mu_r = \{1\}$ (the last equality holds because r is odd). $\qquad\square$

Definition 11.3 Let L be a number field, let $\mathfrak{m} \neq 0$ be an integral ideal of L, and let $I_\mathfrak{m}$ be the group of fractional ideals of L which are prime to $\mathfrak{m}$. Then a group homomorphism $\varphi : I_\mathfrak{m} \to \mathbb{C}^\times$ is called a *Hecke character of L with module of definition* $\mathfrak{m}$ if there are integers n_σ for all $\sigma \in \mathrm{Hom}_\mathbb{Q}(L, \mathbb{C})$, such that for all principal ideals $(a) \in I_\mathfrak{m}$ with $a \equiv 1 \pmod{\mathfrak{m}}$ and a totally positive, the following holds: $\varphi((a)) = \prod_\sigma(\sigma a)^{n_\sigma}$. In this case, the tuple $(n_\sigma)_\sigma$ is called the *infinity type* of φ.

Proposition 11.1 *Let $\mathfrak{p}$ be a prime ideal of $\mathbb{Q}(\zeta)$ which is prime to r, and let $\kappa_\mathfrak{p}$ be the residue field of $\mathfrak{p}$. Then $\kappa_\mathfrak{p}$ is a finite field that contains the r-th roots of unity, and we have a natural isomorphism between the r-th roots of unity in $\mathbb{Q}(\zeta)$ and those in $\kappa_\mathfrak{p}$ which induces a natural embedding $\mu_r \hookrightarrow \mathbb{Q}(\zeta)$ in this case. With this embedding, the character $\chi : \kappa^\times \twoheadrightarrow \mu_r \hookrightarrow \mathbb{Q}(\zeta)$ is given as follows:*

$$\forall x \in \kappa^\times \; : \; x^{\frac{|\kappa|-1}{r}} \equiv \chi(x) \pmod{\mathfrak{p}}.$$

For $\boldsymbol{a} \in A_n^r$ put $\mathbf{J}_{\boldsymbol{a}}(\mathfrak{p}) := \mathcal{J}_{\kappa_\mathfrak{p}}^r(\boldsymbol{a})$. Since the group $I_{(r)} = I_{(r^2)}$ is freely generated by primes $\mathfrak{p}$ as above, we get an induced group homomorphism

$$\mathbf{J}_{\boldsymbol{a}} : I_{(r^2)} \longrightarrow \mathbb{C}^\times.$$

Then we have: $\mathbf{J}_{\boldsymbol{a}}$ is a Hecke character of $\mathbb{Q}(\zeta)$ with module of definition (r^2).

Proof. See [Wei52]. □

Lemma 11.5 *Assume $n \geq 4$, and let s, t be two natural numbers with $s, t \geq 3$ and $s + t = n + 2$. Put*

$$A_{s,t}^r := \Big\{ (\boldsymbol{b}, \boldsymbol{c}) \in A_s^r \times A_t^r \,|\, b_s + c_t = 0 \Big\},$$

and define the maps

$$A_{s,t}^r \xrightarrow{\#} A_n^r, (\boldsymbol{b}, \boldsymbol{c}) \mapsto \boldsymbol{b} \# \boldsymbol{c} := (b_1, \dots, b_{s-1}, c_1, \dots, c_{t-1}) \text{ and}$$

$$A_{s-1}^r \times A_{t-1}^r \xrightarrow{*} A_n^r, (\boldsymbol{b}', \boldsymbol{c}') \mapsto \boldsymbol{b}' * \boldsymbol{c}' := (b_1', \dots, b_{s-1}', c_1', \dots, c_{t-1}').$$

Then the map $A_{s,t}^r \coprod \big(A_{s-1}^r \times A_{t-1}^r\big) \xrightarrow{\# \sqcup *} A_n^r$ is a bijection.

Proof.   This easy lemma can be found in [Shi79]. For the convenience of the reader, we give the inverse map:

$$A_n^r \longrightarrow A_{s,t}^r \coprod \left(A_{s-1}^r \times A_{t-1}^r\right)$$

$$(a_i)_i \mapsto \begin{cases} \underbrace{\left((a_1, \ldots, a_{s-1}), (a_s, \ldots, a_n)\right)}_{\in A_{s-1}^r \times A_{t-1}^r} & \text{if } \sum_{i=1}^{s-1} a_i = 0 \\[2em] \underbrace{\left((a_1, \ldots, a_{s-1}, -\sum_{i=1}^{s-1} a_i), (a_s, \ldots, a_n, -\sum_{i=s}^{n} a_n)\right)}_{\in A_{s,t}^r} & \text{otherwise.} \end{cases}$$

$\square$

Proposition 11.2   *Assume* $q \equiv 1 \pmod{r}$ *(which means that* k *contains the* r*-th roots of unity) Then we have the following rules for Jacobi sums over* k:

(1) Let $\boldsymbol{a} \in A_n^r$ *and* $t \in (\mathbb{Z}/r\mathbb{Z})^\times$ *be arbitrary elements, and let* σ *be the automorphism of* $\mathbb{Q}(\zeta)$ *defined by* $\zeta \mapsto \zeta^t$. *Then*

$$\mathcal{J}(t\boldsymbol{a}) = \sigma\left(\mathcal{J}(\boldsymbol{a})\right).$$

(2) Let $\psi : (k, +) \longrightarrow \mathbb{C}^\times$ *be the character of the additive group of* k *defined by* $\psi(x) := \exp\left(\frac{\mathrm{Tr}_{k/\mathbb{F}_p}(x)}{p}\right)$, *and let* $\xi : k^\times \longrightarrow \mathbb{C}^\times$ *be a non-trivial character of the multiplicative group* $k^\times$. *Define the* Gaussian sum

$$G(\xi) := \sum_{x \in k^\times} \xi(x) \cdot \psi(x).$$

Then $G(\xi) \cdot \overline{G(\xi)} = q$ *and* $G(\bar{\xi}) = \xi(-1) \cdot \overline{G(\xi)}$, *and for* $\boldsymbol{a} \in A_n^r$, *we have the product formula*

$$\mathcal{J}(\boldsymbol{a}) = \frac{(-1)^n}{q} \cdot G(\chi^{a_1}) \cdot \ldots \cdot G(\chi^{a_n}).$$

In particular, we see that $\mathcal{J}(\boldsymbol{a})$ *only depends on* $[\boldsymbol{a}]$.

(3) Let $n \geq 4$, *and let* $s, t \in \mathbb{N}$ *with* $s, t \geq 3$ *and* $s + t = n + 2$. *Then for* $(\boldsymbol{b}, \boldsymbol{c}) \in A_{s,t}^r$ *and* $(\boldsymbol{b}', \boldsymbol{c}') \in A_{s-1}^r \times A_{t-1}^r$ *we have:*

$$\mathcal{J}(\boldsymbol{b} \# \boldsymbol{c}) = \chi(-1)^{b_s} \cdot \mathcal{J}(\boldsymbol{b}) \cdot \mathcal{J}(\boldsymbol{c}) \tag{11.9}$$

and

$$J(\boldsymbol{b'} * \boldsymbol{c'}) = q \cdot J(\boldsymbol{b'}) \cdot J(\boldsymbol{c'}). \tag{11.10}$$

Proof. Compare [Wei49] and [GY95, pp13].

Gouvêa and Yui show (1), Weil proves $G(\xi) \cdot \overline{G(\xi)} = q$ and the product formula from (2). One easily calculates:

$$
\begin{aligned}
G(\bar\xi) &= \sum_{x \in k^\times} \overline{\xi(x)} \cdot \psi(x) = \sum_{x \in k^\times} \overline{\xi(-x)} \cdot \psi(-x) \\
&= \sum_{x \in k^\times} \overline{\xi(-1)} \cdot \overline{\xi(x)} \cdot \overline{\psi(x)} \\
&= \xi(-1) \sum_{x \in k^\times} \overline{\xi(x) \cdot \psi(x)} = \xi(-1) \cdot \overline{G(\xi)}.
\end{aligned}
$$

Formula (11.9) is stated differently in [GY95], so we want to give a proof of (3) here:

$$
\begin{aligned}
J(\boldsymbol{b} \sharp \boldsymbol{c}) &\overset{(2)}{=} \tfrac{(-1)^n}{q} \cdot G(\chi^{b_1}) \cdot \ldots \cdot G(\chi^{b_{s-1}}) \cdot G(\chi^{c_1}) \cdot \ldots \cdot G(\chi^{c_{t-1}}) \\
&= \tfrac{(-1)^n}{q} \cdot \left[(-1)^s \cdot q \cdot G(\chi^{b_s})^{-1} J(\boldsymbol{b}) \right] \cdot \left[(-1)^t \cdot q \cdot G(\chi^{c_t})^{-1} J(\boldsymbol{c}) \right] \\
&= \underbrace{(-1)^{n+s+t}}_{=(-1)^{2n+2}=1} \cdot J(\boldsymbol{b}) \cdot J(\boldsymbol{c}) \cdot q / [\underbrace{G(\chi^{b_s}) \cdot G(\overline{\chi^{b_s}})}_{\substack{\overset{(2)}{=} G(\chi^{b_s}) \cdot \chi(-1)^{b_s} \cdot \overline{G(\chi^{b_s})} \\ \overset{(2)}{=} q \chi(-1)^{b_s}}}], \\
&= \chi(-1)^{b_s} \cdot J(\boldsymbol{b}) \cdot J(\boldsymbol{c})
\end{aligned}
$$

$$
\begin{aligned}
J(\boldsymbol{b'} * \boldsymbol{c'}) &\overset{(2)}{=} \tfrac{(-1)^n}{q} \cdot G(\chi^{b'_1}) \cdot \ldots \cdot G(\chi^{b'_{s-1}}) \cdot G(\chi^{c'_1}) \cdot \ldots \cdot G(\chi^{c'_{t-1}}) \\
&= \tfrac{(-1)^n}{q} \cdot \left[(-1)^{s-1} \cdot q \cdot J(\boldsymbol{b'}) \right] \cdot \left[(-1)^{t-1} \cdot q \cdot J(\boldsymbol{c'}) \right] \\
&= \underbrace{(-1)^{n+s+t-2}}_{=(-1)^{2n}=1} \cdot q \cdot J(\boldsymbol{b'}) \cdot J(\boldsymbol{c'}) \\
&= q \cdot J(\boldsymbol{b'}) \cdot J(\boldsymbol{c'}). \qquad \square
\end{aligned}
$$

Example 11.1 Consider the special case $n = 6$, $r = 3$ and $p = q = 7$.

Put $\boldsymbol{a} := (1,1,1,2,2,2) \in A_6^3$ and $\boldsymbol{b} := (1,1,1,1,1,1) \in A_6^3$. One easily sees that with this, we have $A_6^3 = [\boldsymbol{a}] \sqcup \{\boldsymbol{b}\} \sqcup \{2\boldsymbol{b}\}$. Then $\mu_3 = \{1,2,4\} \subseteq \mathbb{F}_7^\times$, so an embedding $\iota : \mu_3 \hookrightarrow \mathbb{Q}(\zeta)$ is given by sending 2 to ζ, and $\chi : \mathbb{F}_7^\times \twoheadrightarrow \mu_3 \hookrightarrow \mathbb{Q}(\zeta)$ is given by

$$\mathbb{F}_7^\times \longrightarrow \mu_3 \overset{\iota}{\hookrightarrow} \mathbb{Q}(\zeta)$$

$$x \longmapsto x^{\frac{7-1}{3}} = x^2$$

$$2 \longmapsto \zeta.$$

We want to compute $\mathcal{J}(a)$, $\mathcal{J}(b)$ and $\mathcal{J}(2b)$. First we see that $(2, 1, 2, 1, 2, 1) \in [a]$ and

$$(2, 1, 2, 1, 2, 1) = (2, 1, 2, 1) \# (1, 2, 1, 2) = [(2, 1) * (2, 1)] \# [(1, 2) * (1, 2)],$$

and we have

$$\mathcal{J}(1, 2) = \mathcal{J}(2, 1) \overset{\text{Lemma 11.4}}{=} 1.$$

This implies

$$\mathcal{J}(a) \overset{\text{proposition 11.2(2)}}{=} \underset{\iota[(-1)^2] = \iota(1) = 1}{\chi(-1)^1} \cdot [7 \cdot 1 \cdot 1] \cdot [7 \cdot 1 \cdot 1] = 49.$$

Since $b = (1, 1, 1) * (1, 1, 1)$, we now first calculate $\mathcal{J}(1, 1, 1)$ (note that $1 + \zeta + \zeta^2 = 0$):

$$\mathcal{J}(1, 1, 1)$$
$$= (-1)^3 \cdot \big[\chi(1)\chi(5) + \chi(2)\chi(4) + \chi(3)\chi(3) + \chi(4)\chi(2) + \chi(5)\chi(1) \big]$$
$$= -\big[\chi(5) + \chi(8) + \chi(9) + \chi(8) + \chi(5) \big] \quad = -\big[2\chi(5) + 2\chi(1) + \chi(2) \big]$$
$$= -\big[2 \cdot \iota(25) + 2 \cdot \iota(1) + \iota(4) \big] \quad = -\big[2 \cdot \iota(4) + 2 \cdot \iota(1) + \iota(4) \big]$$
$$= -\big[2\zeta^2 + 2 + \zeta^2 \big] \quad = -2 - 3\zeta^2 \quad = -2 - 3(-1 - \zeta)$$
$$= 1 + 3\zeta.$$

With this we get

$$\mathcal{J}(b) \overset{(11.10)}{=} 7 \cdot (1 + 3\zeta)^2 = 7 \cdot (1 + 6\zeta + 9\zeta^2) = 7 \cdot (-8 - 3\zeta)$$
$$= -56 - 21\zeta.$$

Finally, we want to compute $\mathcal{J}(2b)$ using proposition 11.2(1). For this we consider the automorphism σ of $\mathbb{Q}(\zeta)$ given by $\zeta \mapsto \zeta^2 = -1 - \zeta$. With this we get

$$\mathcal{J}(2b) = \sigma\big(\mathcal{J}(b)\big) = \sigma(-56 - 21\zeta) = -56 - 21(-1 - \zeta) = -35 + 21\zeta.$$

Now we get to the announced result by Deligne:

Proposition 11.3 *We have*

$$V = \underbrace{\left(\bigoplus_{a \in A_n^r} V_a \right)}_{=: \tilde{V}_n^r =: \tilde{V}} \oplus \begin{cases} 0 & \text{if } n \text{ is odd} \\ \mathbb{Q}_l \left(\frac{2-n}{2} \right) & \text{if } n \text{ is even,} \end{cases}$$

and the V_a are one dimensional $\mathbb{Q}_l$-vector spaces. Since $\mathcal{X}$ is a hypersurface, for $i \in \{0, 1, \ldots, 2n - 4\} \setminus \{n - 2\}$ we have

$$H_{\text{ét}}^i(\bar{\mathcal{X}}, \mathbb{Q}_l) = \begin{cases} 0 & \text{if } i \text{ is odd} \\ \mathbb{Q}_l(-\frac{i}{2}) & \text{if } i \text{ is even.} \end{cases}$$

If in addition to that we have $q \equiv 1 \pmod{r}$ (so that the V_a are $\mathbb{Q}_l$-G_k-representations according to corollary 11.2), then for all $a \in A_n^r$, the geometric Frobenius acts by multiplication by $\mathcal{J}_k^r(a)$ on V_a.

Proof. See [Del82, I., §7] and compare proposition 5.3. □

Lemma 11.6 *Let a be an arbitrary element of $\check{A} \setminus A_n^r$.*

(1) If $a \neq (0, \ldots, 0)$ or if n is odd, then $V_a = 0$.
(2) If $a = (0, \ldots, 0)$ and if n is even, then V_a is one dimensional, we have $F^*(V_a) = V_a$ and $F^*|V_a = q^{\frac{n-2}{2}}$, and for all $s \in \check{A}$ we have $s^*(V_a) = V_a$ and $s^*|V_a = 1_{V_a}$.

Proof. If n is odd, the statement follows immediately from proposition 11.3, so let us assume that n is even. First consider the case $a = (0, \ldots, 0)$.

If $n = 2$, then $V = H_{\text{ét}}^*(\bar{\mathcal{X}}, \mathbb{Q}_l)$ is the full cohomology ring and hence a $\mathbb{Q}_l$-algebra. Let R be the subring $\mathbb{Q}_l \cdot 1_V$ of V (which is of course isomorphic to $\mathbb{Q}_l$), then any $\mathbb{Q}_l$-algebra-endomorphism of V must restrict to the identity on R. Since l-adic cohomology maps endomorphisms of $\bar{\mathcal{X}}$ to $\mathbb{Q}_l$-algebra-endomorphisms of V, the restriction $f^*|R$ must be the identity on R for every endomorphism f of $\bar{\mathcal{X}}$. In particular, this is true for all $f \in \check{A} \sqcup \{F\}$. By definition of V_a, this shows $R \subseteq V_a$. On the other hand, we know from proposition 11.3 that $\dim_{\mathbb{Q}_l} V_{\boldsymbol{a}} \leq 1$. This proves $R = V_{\boldsymbol{a}}$ and $\dim_{\mathbb{Q}_l} V_{\boldsymbol{a}} = 1$.

let $[H]$ be the class of the smooth hyperplane section $\{X_n = 0\}$ in $\mathrm{CH}^1(\bar{\mathcal{X}})$, and let $\gamma := \mathrm{cl}[H] \in \mathrm{H}^2_{\text{ét}}(\bar{\mathcal{X}}, \mathbb{Q}_l(1))$ be the associated class in cohomology. Since $n \geq 4$, the variety $\bar{\mathcal{X}}$ is irreducible, and according to the strong Lefschetz theorem, multiplication by γ^{n-2} is an isomorphism

$$\mathrm{H}^0_{\text{ét}}(\bar{\mathcal{X}}, \mathbb{Q}_l) \xrightarrow[\sim]{\cup\gamma^{n-2}} \mathrm{H}^{2n-4}_{\text{ét}}(\bar{\mathcal{X}}, \mathbb{Q}_l(n-2)),$$

which in particular implies $[\gamma(-1)]^{\frac{n-2}{2}} \neq 0 \in V$. For all $s \in \tilde{A}$, we obviously have $s^*[H] = H$, so $\tilde{A}$ acts trivially on the one dimensionally subspace $W := \left\langle [\gamma(-1)]^{\frac{n-2}{2}} \right\rangle$ of V. In particular, the group μ_r^n/μ_r acts trivially on W, so $W \subseteq V_{\boldsymbol{a}}$ and hence $W = V_{\boldsymbol{a}}$ since

$$1 = \dim_{\mathbb{Q}_l} W \leq \dim_{\mathbb{Q}_l} V_{\boldsymbol{a}} \overset{\text{proposition 11.3}}{\leq} 1.$$

To compute the restriction of F^* to $V_{\boldsymbol{a}}$, we note that $F^*[H] = [X_n^q = 0] = q[X_n = 0] = qH$, so

$$F^* \left[\gamma^{\frac{n-2}{2}} \right] = \left[F^*\gamma \right]^{\frac{n-2}{2}} = \left[F^*(\mathrm{cl}[H]) \right]^{\frac{n-2}{2}} = \left[\mathrm{cl}(F^*[H]) \right]^{\frac{n-2}{2}}$$

$$= \left[\mathrm{cl}(q[H]) \right]^{\frac{n-2}{2}} = \left[q \cdot \mathrm{cl}[H] \right]^{\frac{n-2}{2}} = q^{\frac{n-2}{2}} \cdot \gamma^{\frac{n-2}{2}}.$$

This proves the lemma for $\boldsymbol{a} = (0, \dots, 0)$. But then it follows trivially from proposition 11.3 that the lemma is true for other $\boldsymbol{a} \in \check{A} \setminus A_n^r$ as well. $\quad\square$

The case of k *not* containing μ_r is not considered in any of the papers by Weil, Deligne, Shioda or Gouvêa/ Yui mentioned in the introduction. We already got a partial result for that case in corollary 11.1; now we are going to study the Frobenius action on the G_k-invariant subspaces $V_{\langle\boldsymbol{a}\rangle}$ in more detail:

Proposition 11.4 *Let a be an arbitrary element of $a \in A_n^r$. Define*

$$
\begin{aligned}
d &:= gcd(r, a_1, \ldots, a_n) && \in \mathbb{N}_+, \\
r' &:= \frac{r}{d} && \in \mathbb{N}_+, \\
e &:= order\ of\ q\ in\ (\mathbb{Z}/r'\mathbb{Z})^\times && \in \mathbb{N}_+, \\
a' &:= \left(\frac{a_1}{d}, \ldots, \frac{a_n}{d}\right) && \in A_n^{r'}.
\end{aligned}
$$

Then $\mathbb{F}_{q^e}$ is the smallest extension of k that contains the r'-th roots of unity, and we have $\#\langle a \rangle = e$.

Let v be an arbitrary vector in $V_a \setminus \{0\}$, and put $v_i := (F^*)^i v \in V_{q^i a} \setminus \{0\}$ for $i \in \{0, \ldots, e-1\}$. Then $\{v = v_0, v_1, \ldots, v_{e-1}\}$ is a basis of $V_{\langle a \rangle}$, and with respect to this basis, the matrix of $F^*|V_{\langle a \rangle}$ is

$$
\left(
\begin{array}{ccc|c}
0 & \cdots & 0 & \mathcal{J}_{\mathbb{F}_{q^e}}^{r'}(a') \\
\hline
1 & & 0 & 0 \\
 & \ddots & & \vdots \\
0 & & 1 & 0
\end{array}
\right).
\tag{11.11}
$$

Proof. For $e' \in \mathbb{N}_0$ we have

$$
\left[\mu_r \subseteq \mathbb{F}_{q^{e'}}\right] \iff \left[q^{e'} \equiv 1 \pmod{r}\right],
$$

so by definition of e, the field $\mathbb{F}_{q^e}$ is really the smallest extension of k that contains μ_r.

Next we want to check that $\#\langle a \rangle = e$. Let $\tilde{e} := \#\langle a \rangle$. Then we obviously have

$$
\langle a \rangle = \{a, qa, q^2 a, \ldots, q^{\tilde{e}-1} a\}
$$

and $q^{\tilde{e}} a = a$. By definition of d, there are $b, b_1, \ldots, b_n \in \mathbb{Z}$ with

$$
br + b_1 a_1 + \ldots + b_n a_n = d,
$$

so we get

$$
q^{\tilde{e}} d \equiv q^{\tilde{e}} b_1 a_1 + \ldots + q^{\tilde{e}} b_n a_n \equiv b_1 a_1 + \ldots + b_n a_n \equiv d \pmod{r}.
$$

This means that r divides $\left[q^{\tilde{e}} - 1\right] d$ and $r' = \frac{r}{d}$ divides $q^{\tilde{e}} - 1$, so $q^{\tilde{e}} \equiv 1$ (mod r'), which shows that e divides $\tilde{e}$ by definition of e.

On the other hand, we have $q^e(a_j/d) \equiv (a_j/d)$ (mod r') for all $i \in \{1,\ldots,n\}$ and hence $q^e a_j \equiv a_j$ (mod r) for all j, so $q^e \boldsymbol{a} = \boldsymbol{a}$, and $\tilde{e}$ divides e. It follows that $e = \tilde{e}$ and that $\{v_0,\ldots,v_{e-1}\}$ is a basis of $V_{\langle\boldsymbol{a}\rangle}$.

In order to prove that F^* is given by (11.11) on $V_{\langle\boldsymbol{a}\rangle}$, we obviously have to show that $(F^*)^e v = \mathcal{J}_{\mathbb{F}_{q^e}}^{r'}(\boldsymbol{a}')v$ holds. For that, we replace k by $k' := \mathbb{F}_{q^e}$ and consider $\mathcal{X} \times_k k'$ instead of $\mathcal{X}$. Then the geometric Frobenius F' on $\bar{\mathcal{X}}$ with respect to k' is F^e, and by lemma 11.3 we get a modified left-$(A \rtimes \mathbb{Z})$-action on $\bar{\mathcal{X}}$ (where $(a \cdot c)$ now acts as $a \circ f^{ec}$). Put $B := \mu_{r'}^n/\mu_{r'}$, and consider the epimorphism $\varphi : A \twoheadrightarrow B$, $(\zeta_i) \mapsto (\zeta_i^d)$. Note that φ is $\mathbb{Z}$-equivariant with respect to the $\mathbb{Z}$-actions on A and B from lemma 11.3. Let $\mathcal{Y} := \mathcal{X}_n^{r'}$ (over k') and $\bar{\mathcal{Y}} := \mathcal{Y} \times_k K$, denote the geometric Frobenius on $\bar{\mathcal{Y}}$ (with respect to k') by F', and consider the left-$(B \rtimes \mathbb{Z})$-action on $\bar{\mathcal{Y}}$ defined in lemma 11.3 (applied to r' instead of r).

Now look at the following morphism of K-varieties:

$$g : \bar{\mathcal{X}} \longrightarrow \bar{\mathcal{Y}}, \quad [x_1 : \ldots : x_n] \mapsto [x_1^d, \ldots, x_n^d].$$

The morphism is obviously dominant and finite, so that the induced $\mathbb{Q}_l$-linear map $g^* : \mathrm{H}^{n-2}_{\mathrm{\acute{e}t}}(\bar{\mathcal{Y}}, \mathbb{Q}_l) \to \mathrm{H}^{n-2}_{\mathrm{\acute{e}t}}(\bar{\mathcal{X}}, \mathbb{Q}_l)$ is *injective*. We claim that g^* is $A \rtimes \mathbb{Z}$-equivariant. To see that, first note that for arbitrary $a = (\zeta_i) \in A$ and an arbitrary K-rational point $[x_1,\ldots,x_n]$ of $\bar{\mathcal{X}}$, we have

$$\left(g \circ a\right)[x_1 : \ldots : x_n] = g\left[\zeta_1 x_1 : \ldots : \zeta_n x_n\right] = \left[\zeta_1^d x_1^d : \zeta_n^d x_n^d q\right]$$

$$= \varphi(a)\left[x_1^d : \ldots : x_n^d q\right] = \left(\varphi(a) \circ g\right)[x_1 : \ldots : x_n].$$

This means that for all $a \in A$, the following diagram commutes:

$$\begin{array}{ccc}
\mathrm{H}^{n-2}_{\mathrm{\acute{e}t}}(\bar{\mathcal{Y}}, \mathbb{Q}_l) & \xrightarrow{\;g^*\;} & \mathrm{H}^{n-2}_{\mathrm{\acute{e}t}}(\bar{\mathcal{X}}, \mathbb{Q}_l) \\
{\scriptstyle\varphi(a)^*}\downarrow{\scriptstyle\wr} & & {\scriptstyle\wr}\downarrow{\scriptstyle a^*} \\
\mathrm{H}^{n-2}_{\mathrm{\acute{e}t}}(\bar{\mathcal{Y}}, \mathbb{Q}_l) & \xrightarrow[\;g^*\;]{} & \mathrm{H}^{n-2}_{\mathrm{\acute{e}t}}(\bar{\mathcal{X}}, \mathbb{Q}_l).
\end{array} \qquad (11.12)$$

On the other hand, it is obvious that $F' \circ g = g \circ F'$ and hence $g^* \circ F'^* =$

$F'^* \circ g^*$. So for all $a \in A$ and $c \in \mathbb{Z}$, we have

$$g^* \circ (a \cdot c)^* = g^* \circ (a \circ f^{ec})^* = g^* \circ (f^*)^{ec} \circ a^* = g^* \circ (F'^*)^{-c} \circ a^*$$
$$= (F'^*)^c \circ g^* \circ a^* \overset{(11.12)}{=} (F'^*)^c \circ \varphi(a)^* \circ g^* = (\varphi(a) \cdot c)^* \circ g^*,$$

which shows that g^* is really $(A \rtimes \mathbb{Z})$-equivariant.

Again, let $\mathcal{M}$ be the category of $\mathbb{Q}_l$-vector spaces , and consider the objects $M := H^{n-2}_{\text{ét}}(\bar{\mathcal{Y}}, \mathbb{Q}_l)$ and $N := H^{n-2}_{\text{ét}}(\bar{\mathcal{X}}, \mathbb{Q}_l)$ of $\mathcal{M}$. In addition to that, let $B := A^{r'}_n$, and let $\chi := a' \in \check{B}$. Then for all $a = (\zeta_i) \in A$, we have

$$[\check{\varphi}(\chi)](a) = \chi[\varphi(a)] = a'(\zeta_i^d) = (\zeta_1^d)^{\frac{a_1}{d}} \cdot \ldots \cdot (\zeta_n^d)^{\frac{a_n}{d}}$$
$$= \zeta_1^{a_1} \cdot \ldots \cdot \zeta_n^{a_n} = a(a),$$

so $\check{\varphi}(\chi) = a$. Therefore $M_\chi = \left[H^{n-2}_{\text{ét}}(\bar{\mathcal{Y}}, \mathbb{Q}_l)\right]_{a'}$ and $N_{\check{\varphi}(\chi)} = V_a$, and since $q^e a' = a'$ and $q^e a = a$, both spaces are invariant under F' and hence invariant under the $\mathbb{Z}$-action, so that we also have $M_{[\chi]} = M_\chi$ and $N_{[\check{\varphi}(\chi)]} = N_{\check{\varphi}(\chi)}$ (using the notation of corollary 10.2).

We have seen above that $g^* : M \to N$ is an $(A \rtimes \mathbb{Z})$-equivariant morphism in $\mathcal{M}$, so we can apply corollary 10.2 and get the following commutative diagram:

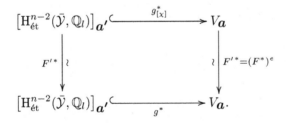

Since k' contains the r'-th roots of unity, we know from proposition 11.3, that F'^* acts as multiplication by $\mathcal{J}^{r'}_{k'}(a')$ on $\left[H^{n-2}_{\text{ét}}(\bar{\mathcal{Y}}, \mathbb{Q}_l)\right]_{a'}$. In addition to that, both spaces $\left[H^{n-2}_{\text{ét}}(\bar{\mathcal{Y}}, \mathbb{Q}_l)\right]_{a'}$ and V_a are one dimensional, so the injection $g^*_{[\chi]}$ must be an isomorphism. This means that we have the following commutative diagram:

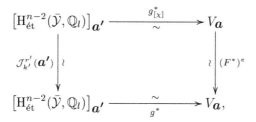

and the proposition is proven.   □

Example 11.2   Consider the special case $r := n := 3$ and $q := p := 5$
(note that $k = \mathbb{F}_5$ does *not* contain the third roots of unity).

We see immediately that $A_3^3 = \{(1,1,1),(2,2,2)\}$, so we have $\bar{V} = V_{\langle(1,1,1)\rangle}$
since $q \equiv 2 \pmod 3$. We want to apply proposition 11.4 to $\boldsymbol{a} := (1,1,1)$, we
find $d = 1$, $r' = r = 3$, $e = 2$ and $\boldsymbol{a'} = \boldsymbol{a} = (1,1,1)$, and in the appendix we
compute the necessary Jacobi sum, namely $\mathcal{J}_{\mathbb{F}_{q^e}}^{r'}(\boldsymbol{a'}) = \mathcal{J}_{\mathbb{F}_{25}}^3(1,1,1) = -5$.

Hence we see that there are cohomology classes $v_0 \in V_{(1,1,1)}$ and $v_1 \in$
$V_{(2,2,2)}$ that form a basis of $\bar{V}$, such that the matrix of $F^*|\bar{V}$ with respect
to this basis is $\left(\begin{smallmatrix} 0 & -5 \\ 1 & 0 \end{smallmatrix}\right)$.

We want to understand the action of $\tilde{A}$ on $V \cong V_{(0,\ldots,0)} \oplus \bar{V}$. We know
from lemma 11.6, that $\tilde{A}$ acts trivially on $V_{(0,\ldots,0)}$, and example 10.4(2)
describes the action of the normal subgroup μ_r^n/μ_r of $\tilde{A}$ on $\bar{V}$. Hence we
"only" have to understand how S_n acts on $\bar{V}$. We start by studying the
action of a transposition on $\bar{V}$.

Lemma 11.7   *Let $\tau \in S_n$ be a transposition, and let $\boldsymbol{a} \in A_n^r$ be an
arbitrary element satisfying ${}^{\tau}\boldsymbol{a} \neq \boldsymbol{a}$. Then τ^* induces a $\mathbb{Q}_l$-linear involution
of $V_{\boldsymbol{a}} \oplus V_{\tau\boldsymbol{a}}$ whose trace is zero.*

Proof.   By corollary 10.1 and because of ${}^{\tau}({}^{\tau}\boldsymbol{a}) = \boldsymbol{a}$, the map τ^* restricts
to $\mathbb{Q}_l$-linear maps $V_{\boldsymbol{a}} \xrightarrow{\tau_{\tau_a}^*} V_{\tau\boldsymbol{a}}$ and $V_{\tau\boldsymbol{a}} \xrightarrow{\tau_a^*} V_{\boldsymbol{a}}$, so that we really get an
induced endomorphism $\tilde{\tau}^*: V_{\boldsymbol{a}} \oplus V_{\tau\boldsymbol{a}}$ which must be an involution since
$\tau^2 = 1_{S_n}$.

By proposition 11.3, the subspaces $V_{\boldsymbol{a}}$ and $V_{\tau\boldsymbol{a}}$ have dimension one. Let $v \in V_{\boldsymbol{a}} \setminus \{0\}$ be an arbitrary cohomology class. Then $\tau(v) \in V_{\tau\boldsymbol{a}}$ can not be zero, because τ^* is an automorphism of V. It follows that $\{v, \tau^*(v)\}$ is a basis of $V_{\boldsymbol{a}} \oplus V_{\tau\boldsymbol{a}}$.

Since τ^* is an involution, the matrix of $\tilde{\tau}^*$ with respect to this basis is $\left\{\begin{smallmatrix} 0 & 1 \\ 1 & 0 \end{smallmatrix}\right\}$, and this matrix obviously has trace zero. $\qquad\square$

The following proposition is the key result needed for the computation of the S_n-action on $\bar{V}$. Its proof is surprisingly difficult and can be considered as the technical heart of this book.

Proposition 11.5 *Let $\tau \in S_n$ be a transposition, and let $\boldsymbol{a} \in A_n^r$ be an arbitrary element satisfying $^\tau\boldsymbol{a} = \boldsymbol{a}$. Then τ^* induces an involution of $V_{\boldsymbol{a}}$ which is multiplication by (-1).*

Proof. Since $^\tau\boldsymbol{a} = \boldsymbol{a}$, we know from corollary 10.1 that τ^* restricts to an involution $V_{\boldsymbol{a}} \xrightarrow{\tau_{\boldsymbol{a}}^*} V_{\boldsymbol{a}}$. Since $V_{\boldsymbol{a}}$ has dimension one by proposition 11.3, this automorphism can only be the identity or multiplication by (-1), and we have to prove that it is multiplication by (-1).

Without loss of generality, we can obviously assume that $\tau = [12]$, which means we have $a_1 = a_2$. We are going to prove the proposition separately for the cases $n = 2$, $n = 3$ and $n \geq 4$:

- $\underline{n = 2}$: If r is odd, then there is no $\boldsymbol{a} \in A_2^r$ with $a_1 = a_2$. So r must be even, and we must have $\boldsymbol{a} = (\frac{r}{2}, \frac{r}{2})$.

  In order to show that $\tau_{(\frac{r}{2}, \frac{r}{2})}^*$ is multiplication by (-1), we want to use the Lefschetz trace formula (5.8) to prove that the trace of $\tau_{(\frac{r}{2}, \frac{r}{2})}^*$ is (-1). We claim that the automorphism τ of $\bar{\mathcal{X}}$ has no fixed points: We have $[0:1], [1,0] \notin \bar{\mathcal{X}}(K)$, and if $[1:y]$ was a fixed point with $y \in K^\times$, then $[1:y] = [y:1]$ and hence $y^2 = 1$. But then

  $$1^r + y^r = 1 + (y^2)^{\frac{r}{2}} = 2 \overset{p > r \geq 3}{\neq} 0,$$

  a contradiction to $[1:y] \in \bar{\mathcal{X}}(K)$.

It is

$$V = V_{(0,0)} \oplus V_{(\frac{r}{2},\frac{r}{2})} \oplus \underbrace{\bigoplus_{0<j<\frac{r}{2}} \left(V_{(j,r-j)} \oplus V_{(r-j,j)} \right)}_{=:V'}.$$

According to lemma 11.6, $V_{(0,0)}$ is one dimensional, and the restriction of τ^* to $V_{(0,0)}$ is the identity and therefore has trace one. Furthermore, lemma 11.7 implies that the trace of τ^* restricted to V' is zero. So by the Lefschetz trace formula, we get:

$$0 = \mathrm{Tr}\,(\tau^*) = 1 + \mathrm{Tr}\,\left(\tau^*_{(\frac{r}{2},\frac{r}{2})} \right) + 0 \implies \mathrm{Tr}\,\left(\tau^*_{(\frac{r}{2},\frac{r}{2})} \right) = -1,$$

which proves the proposition in this case.

- $\underline{n=3}$: Again, we first compute the number of fixed points of τ, i.e. the intersection number (Δ, Γ_τ): Let $[x : y : z] \in \bar{\mathcal{X}}(K)$ be such a fixed point, so that $[x : y : z] = [y : x : z]$. We can assume $x \in \{0,1\}$ and consider three cases:

  - 1. case: $\underline{x = 0}$
    Then $y = 0$ and $z \neq 0$, so $x^r + y^r + z^r = 0 + 0 + z^r \neq 0$ — contradiction!
  - 2. case: $\underline{x = 1, y = 1}$
    Then $1^r + 1^r + z^r = 0$, hence $z^r = -2 \neq 0$ (since $p > 2$). It follows that there are r distinct fixed points of this type, one for each of the r distinct r-th roots of (-2) in K.
  - 3. case: $\underline{x = 1, y \notin \{0,1\}}$
    Then

$$[1 : y : z] = [y : 1 : z] \implies [y : y^2 : yz] = [y : 1 : z]$$

$$\implies yz = z \overset{y \neq 1}{\implies} z = 0,\ y = -1 \overset{p>2}{\neq} 1,$$

hence $[x : y : z] = [1 : -1 : 0]$. If r is even, then

$$1^r + (-1)^r + 0^r = 2 \neq 0 \text{ — contradiction!}$$

If on the other hand r is odd, then $[1 : -1 : 0]$ is indeed a K-rational point of $\bar{\mathcal{X}}$ and therefore a fixed point of τ.

So the automorphism τ has r fixed points if m is even and $(r+1)$ fixed points if r is odd. Even if some of these fixed points have multiplicity greater than one in the intersection cycle $\Delta \cdot \Gamma_\tau$, we still get the estimates

$$(\Delta \cdot \Gamma_\tau) \geq \left\{ \begin{matrix} r \\ r+1 \end{matrix} \right\} \quad \text{for} \quad \left\{ \begin{matrix} r \text{ even} \\ r \text{ odd} \end{matrix} \right\}. \qquad (11.13)$$

By the Lefschetz trace formula (5.8) we have:

$$(\Delta_{\bar{\mathcal{X}}}, \Gamma_\tau)$$
$$= \underbrace{\mathrm{Tr}\left[\tau^* \Big| \mathrm{H}^0_{\mathrm{\acute{e}t}}(\bar{\mathcal{X}}, \mathbb{Q}_l)\right]}_{=:t_0} - \underbrace{\mathrm{Tr}\left[\tau^* \Big| \mathrm{H}^1_{\mathrm{\acute{e}t}}(\bar{\mathcal{X}}, \mathbb{Q}_l)\right]}_{=:t_1} + \underbrace{\mathrm{Tr}\left[\tau^* \Big| \mathrm{H}^2_{\mathrm{\acute{e}t}}(\bar{\mathcal{X}}, \mathbb{Q}_l)\right]}_{=:t_2}.$$
$$(11.14)$$

Since $\dim \bar{\mathcal{X}} = 1$, $\bar{\mathcal{X}}$ is irreducible, and $\mathrm{H}^0_{\mathrm{\acute{e}t}}(\bar{\mathcal{X}}, \mathbb{Q}_l)$ is isomorphic to $\mathbb{Q}_l$ (as a $\mathbb{Q}_l$-algebra). Since τ^* is an $\mathbb{Q}_l$-algebra automorphism, its restriction to $\mathrm{H}^0_{\mathrm{\acute{e}t}}(\bar{\mathcal{X}}, \mathbb{Q}_l)$ must be the identity, so that $t_0 = 1$. Since $\mathrm{H}^2_{\mathrm{\acute{e}t}}(\bar{\mathcal{X}}, \mathbb{Q}_l)$ is generated by the class of a smooth hyperplane section (on which τ^* obviously acts trivially), we also have $t_2 = 1$. So equation (11.14) implies:

$$t_1 = 2 - (\Delta_{\bar{\mathcal{X}}}, \Gamma_\tau). \qquad (11.15)$$

Now we ask how many elements $\boldsymbol{b} \in A_3^r$ satisfy $b_1 = b_2$ — denote this number by N:

- 1. case: r odd
  Then $b_1 + b_2 = 2b_1 \not\equiv 0 \pmod{r}$ for all $b_1 \in \mathbb{Z}/r\mathbb{Z} \setminus \{0\}$, so that

$$N = \#\Big\{ (b_1, b_1, -2b_1) \,\Big|\, b_1 \in \mathbb{Z}/r\mathbb{Z} \setminus \{0\} \Big\} = r - 1.$$

- 2. case: r even
  In this case, we must have $b_1 \not\equiv \frac{r}{2} \pmod{r}$, because b_3 would be zero otherwise. Hence in this case, we have

$$N = \#\Big\{ (b_1, b_1, -2b_1) \,\Big|\, b_1 \in \mathbb{Z}/r\mathbb{Z} \setminus \Big\{0, \frac{r}{2}\Big\} \Big\} = r - 2.$$

We have

$$t_1 = \mathrm{Tr}\,(V) \stackrel{\mathrm{proposition 11.3}}{=} \mathrm{Tr}\left(\tau^* \left| \bigoplus_{b \in A_n^r} V_b \right.\right)$$

$$= \sum_{(b_1,b_1,b_3) \in A_n^r} \underbrace{\mathrm{Tr}\left(\tau^* \left| V_{(b_1,b_1,b_3)}\right.\right)}_{\in \{-1,1\}}$$

$$+ \sum_{\substack{(b_1,b_2,b_3) \in A_n^r \\ b_1 < b_2}} \underbrace{\mathrm{Tr}\left(\tau^* \left| V_{(b_1,b_2,b_3)} \oplus V_{(b_2,b_1,b_3)}\right.\right)}_{\stackrel{\mathrm{lemma 11.7}}{=} 0},$$

which implies $t_1 \in [-N, N]$ with $t_1 = -N$ if and only if the proposition is true for $n = 3$.

– 1. case: r odd

$$t_1 \stackrel{(11.15)}{=} 2 - (\Delta_{\bar{\mathcal{X}}}, \Gamma_\tau) \stackrel{(11.13)}{\le} 2 - (r+1) = 1 - r = -N.$$

– 2. case: r even

$$t_1 \stackrel{(11.15)}{=} 2 - (\Delta_{\bar{\mathcal{X}}}, \Gamma_\tau) \stackrel{(11.13)}{\le} 2 - r = -N.$$

Therefore $t_1 = -N$ in any case, and the proposition is proven for the case $n = 3$.

- $n \ge 4$: So now $\mathcal{X}$ has at least dimension two, and we can not use the Lefschetz trace formula as in the first two cases, because Δ and Γ_τ will not intersect properly anymore.

Let s and t be natural numbers satisfying $s, t \ge 3$ and $s + t = n + 2$. We start by defining A-actions on the $\mathbb{Q}_l$-vector spaces $\bar{V}_{s-1}^r \otimes_{\mathbb{Q}_l} \bar{V}_{t-1}^r$ and $[\bar{V}_s^r \otimes_{\mathbb{Q}_l} \bar{V}_t^r]^{\mu_r}$: The map

$$\begin{array}{ccc} A & \longrightarrow & (\mu_r^{s-1}) \times (\mu_r^{t-1}) \\ (\zeta_1, ..., \zeta_n) & \mapsto & \left((\zeta_1, ..., \zeta_{s-1}), (\zeta_s, ..., \zeta_n)\right) \end{array}$$

is obviously a well defined group homomorphism, and composition with the natural action $(\mu_r^{s-1}/\mu_r) \times (\mu_r^{t-1}/\mu_r) \to \bar{V}_{s-1}^r \otimes \bar{V}_{t-1}^r$, gives the desired A-action on $\bar{V}_{s-1}^r \otimes \bar{V}_{t-1}^r$. Next, look at the following sequence:

$$1 \longrightarrow \mu_r \longrightarrow (\mu_r^s/\mu_r) \times (\mu_r^t/\mu_r) \longrightarrow A \longrightarrow 1$$

$$\zeta \longmapsto ((\zeta,...,\zeta,1), (\zeta,...,\zeta,1))$$

$$((\zeta_1,...,\zeta_s), (\xi_1,...,\xi_t)) \longmapsto \left(\frac{\zeta_1}{\zeta_s},...,\frac{\zeta_{s-1}}{\zeta_s}, \frac{\xi_1}{\xi_t},...,\frac{\xi_{t-1}}{\xi_t} \right).$$

It is easy to see that the morphisms are well defined and that the sequence is exact. Starting with the natural action of $(\mu_r^s/\mu_r) \times (\mu_r^t/\mu_r)$ on $\bar{V}_s^r \otimes \bar{V}_t^r$, we therefore get an induced action of A on the subspace of invariants under the kernel μ_r.

Having defined these A-actions, we now look at the following commuting diagram of K-varieties described in [Shi79, p179]:

$$
\begin{array}{ccc}
\beta^{-1}(Y) & \overset{j'}{\hookrightarrow} & Z_{s,t}^r \\
\beta' \downarrow & \square & \beta \downarrow \quad \searrow^{\psi} \\
\bar{\mathcal{X}}_{s-1}^r \times \bar{\mathcal{X}}_{t-1}^r & \overset{\hookrightarrow}{j} & \bar{\mathcal{X}}_s^r \times \bar{\mathcal{X}}_t^r \cdots\overset{}{\underset{\varphi}{\dashrightarrow}} \bar{\mathcal{X}}_n^r
\end{array}
$$

Here j, β, β', j' and ψ are morphism, φ is a rational map, and they are defined as follows:

$$\varphi \quad : \quad ([x_1 : \ldots : x_s], [y_1 : \ldots : y_t]) \mapsto$$
$$[x_1 y_t : \ldots : x_{s-1} y_t : \varepsilon x_s y_1 : \ldots : \varepsilon x_s y_{t-1}] \text{ (with } \varepsilon^r = -1),$$

$$j \quad : \quad ([x_1 : \ldots : x_{s-1}], [y_1 : \ldots : y_{t-1}]) \mapsto$$
$$([x_1 : \ldots : x_{s-1}, 0], [y_1 : \ldots : y_{t-1}, 0]),$$

$$\beta \quad : \quad \text{blow-up of } \bar{\mathcal{X}}_s^r \times \bar{\mathcal{X}}_t^r \text{ in } \bar{\mathcal{X}}_{s-1}^r \times \bar{\mathcal{X}}_{t-1}^r,$$

$$\beta' \quad : \quad \text{restriction of } \beta \text{ to } \beta^{-1}(\bar{\mathcal{X}}_{s-1}^r \times \bar{\mathcal{X}}_{t-1}^r),$$

$$j' \quad : \quad \text{canonical embedding},$$

$$\psi \quad = \quad \varphi \circ \beta.$$

Shioda proves that the morphism

$$\left[\bar{V}_s^r \otimes \bar{V}_t^r\right]^{\mu_r} \oplus \left[\bar{V}_{s-1}^r \otimes \bar{V}_{t-1}^r\right](-1) \xrightarrow[\sim]{\psi_*(\beta^* \oplus j_*' \beta'^*)} \bar{V}_n^r \qquad (11.16)$$

induced by this diagram is an A-equivariant isomorphism of $\mathbb{Q}_l$-G-representations with respect to the A-action on the left-hand side defined above (the "(-1)" denotes a Tate-twist).

If $(\boldsymbol{b}, \boldsymbol{c}) \in A^r_{s,t}$ and $(\boldsymbol{b}', \boldsymbol{c}') \in A^r_{s-1} \times A^r_{t-1}$ are arbitrary elements, then by [Shi79] and [Shi82], equation (11.16) induces isomorphisms

$$V_{\boldsymbol{b}} \otimes V_{\boldsymbol{c}} \xrightarrow[\sim]{\psi_* \beta^*} V_{\boldsymbol{b} \# \boldsymbol{c}} \qquad \text{and} \qquad V_{\boldsymbol{b}'} \otimes V_{\boldsymbol{c}'}(-1) \xrightarrow[\sim]{\psi_* j'_* \beta'^*} V_{\boldsymbol{b}' * \boldsymbol{c}'}.$$

Now consider the special case $s := 3$ and $t := n - 1 (\geq 4 - 1 = 3)$. For $([x_1, x_2, x_3], [y_1, \ldots, y_{n-1}])$ in $\bar{\mathcal{X}}^r_3 \times \bar{\mathcal{X}}^r_{n-1}$, we have

$$\begin{aligned}
\left[\tau \circ \varphi\right] & \left([x_1, x_2, x_3], [y_1, \ldots, y_{n-1}]\right) \\
&= \tau\left([x_1 y_{n-1} : x_2 y_{n-1} : \varepsilon x_3 y_1 : \ldots : \varepsilon x_3 y_{n-2}]\right) \\
&= \left([x_2 y_{n-1} : x_1 y_{n-1} : \varepsilon x_3 y_1 : \ldots : \varepsilon x_3 y_{n-2}]\right) \\
&= \varphi\left([x_2, x_1, x_3], [y_1, \ldots, y_{n-1}]\right) \\
&= \left[\varphi \circ (\tau \times 1)\right]\left([x_1, x_2, x_3], [y_1, \ldots, y_{n-1}]\right),
\end{aligned}$$

so $\tau \circ \varphi = \varphi \circ (\tau \times 1)$, and for $([x_1, x_2], [y_1, \ldots, y_{n-2}])$ in $\bar{\mathcal{X}}^r_2 \times \bar{\mathcal{X}}^r_{n-2}$, we have

$$\begin{aligned}
\left[(\tau \times 1) \circ j\right] & \left([x_1, x_2], [y_1, \ldots, y_{n-2}]\right) \\
&= (\tau \times 1)\left([x_1, x_2, 0], [y_1, \ldots, y_{n-2}, 0]\right) \\
&= \left([x_2, x_1, 0], [y_1, \ldots, y_{n-2}, 0]\right) = j\left([x_2, x_1], [y_1, \ldots, y_{n-2}]\right) \\
&= \left[j \circ (\tau \times 1)\right]\left([x_1, x_2], [y_1, \ldots, y_{n-2}]\right),
\end{aligned}$$

so $(\tau \times 1) \circ j = j \circ (\tau \times 1)$. By the universal property of the blow-up, we therefore get uniquely determined morphisms τ' such that the following diagram commutes:

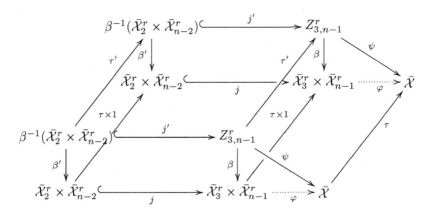

Since τ and τ' are involutions, we have $\tau^* = \tau_*$ and $\tau'^* = \tau'_*$, and using this we get

$$\tau^*\psi_*\beta^* = (\tau_*\psi_*)\beta^* = (\psi_*\tau'_*)\beta^* = \psi_*(\tau'^*\beta^*) = \psi_*\beta^*(\tau \times 1)^*$$

and

$$\begin{aligned}
\tau^*\psi_*j'_*\beta'^* = (\tau_*\psi_*j'_*)\beta'^* &= (\psi_*j'_*\tau'_*)\beta'^* \\
&= \psi_*j'_*(\tau'^*\beta'^*) = \psi_*j'_*\beta'^*(\tau \times 1)^*.
\end{aligned}$$

This implies that for elements $(\boldsymbol{b}, \boldsymbol{c}) \in A^r_{s,t}$ and $(\boldsymbol{b}', \boldsymbol{c}') \in A^r_{s-1} \times A^r_{t-1}$ with $b_1 = b_2$ and $b'_1 = b'_2$, the following two diagrams commute:

$$
\begin{CD}
V_{\boldsymbol{b}} \otimes V_{\boldsymbol{c}} @>{\psi_*\beta^*}>{\sim}> V_{\boldsymbol{b}\#\boldsymbol{c}} \\
@V{(\tau\times 1)^*}V{\wr}V @V{\wr}V{\tau^*}V \\
V_{\boldsymbol{b}} \otimes V_{\boldsymbol{c}} @>{\psi_*\beta^*}>{\sim}> V_{\boldsymbol{b}\#\boldsymbol{c}}
\end{CD}
\qquad
\begin{CD}
V_{\boldsymbol{b}'} \otimes V_{\boldsymbol{c}'}(-1) @>{\psi_*j'_*\beta'^*}>{\sim}> V_{\boldsymbol{b}'*\boldsymbol{c}'} \\
@V{(\tau\times 1)^*}V{\wr}V @V{\wr}V{\tau^*}V \\
V_{\boldsymbol{b}'} \otimes V_{\boldsymbol{c}'}(-1) @>{\psi_*j'_*\beta'^*}>{\sim}> V_{\boldsymbol{b}'*\boldsymbol{c}'}
\end{CD}
$$

According to lemma 11.5, the element $\boldsymbol{a}$ is either of the form $\boldsymbol{b}\#\boldsymbol{c}$ or of the form $\boldsymbol{b}'*\boldsymbol{c}'$ for suitable $\boldsymbol{b}$ and $\boldsymbol{c}$ respectively $\boldsymbol{b}'$ and $\boldsymbol{c}'$ with $b_1 = b_2$ respectively $b'_1 = b'_2$ (since $a_1 = a_2$). Because we have already proved the proposition for "$n = 2$" and "$n = 3$", we know that $(\tau \times 1)^*$ is multiplication by (-1) in either case, and the commutativity of the corresponding diagram shows that τ^* is multiplication by (-1) as well.

So the statement holds for all $n \geq 2$, and the proof of the proposition is finally complete. $\qquad\square$

Now we have enough information to prove the existence of a "nice" basis $\{v_b\}$ of $\bar{V}$ (with $v_b \in V_b$) which allows an explicit description of the action of S_n:

Corollary 11.3   *Let a be an arbitrary element of A_n^r. Then for all elements $b \in [a]$, there are elements $v_b \in V_b \setminus \{0\}$ such that*

$$\forall \sigma \in S_n \ \forall b \in [a] : \ \sigma^*(v_b) = sgn\,(\sigma) \cdot v_{\sigma^{-1}b}.$$

Proof.   First choose an arbitrary $v \in V_a \setminus \{0\}$ and for each $b \in [a]$ a permutation $\sigma_b \in S_n$ with $\sigma_b^{-1}(a) = b$. Put $v_b := sgn\,(\sigma_b) \cdot \sigma_b^* v$. By corollary 10.1, we have $\sigma_b^*(V_a) = V_b$, so $v_b \in V_b \setminus \{0\}$.

Now we will show that v_b is independent of the choice of σ_b: Let $\tilde{\sigma}_b \in S_n$ be another permutation with $\tilde{\sigma}_b^{-1}(a) = b$. This implies that b is fixed by $\omega := \sigma_b^{-1}\tilde{\sigma}_b$. We can write ω as a product of disjoint cycles, and without loss of generality (change the indices if necessary), this product looks like

$$\omega = [1, \ldots, k_1] \cdot [k_1 + 1, \ldots, k_2] \cdot \ldots \cdot [k_{s-1} + 1, \ldots, n].$$

Since b is fixed by ω, we therefore have

$$b_1 = b_2 = \ldots = b_{k_1}, \ b_{k_1+1} = \ldots = b_{k_2}, \ b_{k_{s-1}+1} = \ldots = b_n.$$

So if we write each of the disjoint cycles as a product of transpositions of elements of the cycle, each of these transpositions fixes b as well. This shows that we can write ω as a product of transpositions that all fix b.

Proposition 11.5 then implies that ω^* is multiplication by $sgn\,(\omega)$ on V_b, so

$$sgn\,(\tilde{\sigma}_b) \cdot \tilde{\sigma}_b^* v = sgn\,(\sigma_b \omega) \cdot (\sigma_b \omega)^* v = sgn\,(\sigma_b \omega) \cdot \omega^* \sigma_b^* v$$
$$= sgn\,(\sigma_b \omega) \cdot sgn\,(\omega) \cdot \sigma_b^* v = sgn\,(\sigma_b) \cdot \sigma_b^* v = v_b.$$

Hence we see that v_b is really independent of the choice of σ_b.

Now we are going to use this fact to prove that S_n acts on the v_b as claimed in the corollary. For this, let $\sigma \in S_n$ be an arbitrary permutation, and let $b \in [a]$ be an arbitrary element. We have $(\sigma_b \sigma)^{-1}(a) = \sigma^{-1}(b)$ and therefore by the above independence

$$v_{\sigma^{-1}b} = \text{sgn}\,(\sigma_b \sigma) \cdot (\sigma_b \sigma)^* v.$$

This implies:

$$
\begin{aligned}
\sigma^* v_b = \sigma^*(\text{sgn}\,(\sigma_b) \cdot \sigma_b^* v) &= \text{sgn}\,(\sigma_b) \cdot \sigma^* \sigma_b^* v \\
&= \text{sgn}\,(\sigma_b) \cdot \underbrace{\text{sgn}\,(\sigma) \cdot \text{sgn}\,(\sigma)}_{=1} \cdot (\sigma_b \sigma)^* v \\
&= \text{sgn}\,(\sigma) \cdot \underbrace{\text{sgn}\,(\sigma_b \sigma) \cdot (\sigma_b \sigma)^* v}_{=v_{\sigma^{-1}b}} = \text{sgn}\,(\sigma) \cdot v_{\sigma^{-1}b}.
\end{aligned}
$$

$\square$

Corollary 11.4 *Let $a \in A_n^r$ be an arbitrary element, let $(S_n)_a$ be the subgroup of S_n that fixes a, and let $\textbf{sgn} : (S_n)_a \longrightarrow \mathbb{Q}_l^\times = Aut(V_a)$ be the character mapping a permutation to its sign. Then the action of S_n on $V_{[a]}$ is induced from $\textbf{sgn}$, i.e. we have:*

$$
\begin{bmatrix} S_n \longrightarrow Aut_{\mathbb{Q}_l}(V_{[a]}) \\ \sigma \quad\mapsto\quad (\sigma^{-1})^* \end{bmatrix} \cong \text{ind}_{(S_n)_a}^{S_n}\,[\textbf{sgn}]. \tag{11.17}
$$

In particular, for an arbitrary $\sigma \in S_n$ we have:

$$
\begin{aligned}
\text{Tr}\,(\sigma^* \,|\, V_{[a]}) &= sgn\,(\sigma) \cdot \frac{\#\left\{\tau \in S_n \,\middle|\, \tau\sigma\tau^{-1} \in (S_n)_a\right\}}{\#(S_n)_a} \\
&= sgn\,(\sigma) \cdot \#\left\{b \in [a] \,\middle|\, \sigma \in (S_n)_b\right\}.
\end{aligned}
$$

Proof. Let $\{v_b \in V_b | b \in [a]\}$ be a basis of $V_{[a]}$ as in corollary 11.3. To prove the first statement, we have to show that

$$
\mathbb{Q}_l[S_n] \otimes_{\mathbb{Q}_l[(S_n)_a]} V_a \xrightarrow{\varphi} V_{[a]}, \quad \left(\sum_{\sigma \in S_n} x_\sigma \sigma\right) \otimes v \mapsto \sum_\sigma x_\sigma (\sigma^{-1})^*(v)
$$

(where V_a is considered as a $\mathbb{Q}_l[(S_n)_a]$-module via $\textbf{sgn}$) is a well-defined $\mathbb{Q}_l[S_n]$-linear isomorphism. To see that it is well-defined, the canonical injection $V_a \xrightarrow{\psi} V$ must be shown to be $\mathbb{Q}_l[(S_n)_a]$-linear; since it is clearly $\mathbb{Q}_l$-linear, we only have to show that $\psi(\sigma \cdot v_a) = \sigma \cdot \psi(v_a)$ for all $\sigma \in (S_n)_a$.

This follows from corollary 11.3, because $\sigma \cdot v_a = sgn(\sigma)v_a$.

Now let $\boldsymbol{b}$ be an arbitrary element of $[\boldsymbol{a}]$, and let $\sigma \in S_n$ be a permutation with $^\sigma \boldsymbol{a} = \boldsymbol{b}$. Applying corollary 11.3 again, we see that

$$\varphi(sgn(\sigma)\sigma \otimes v_{\boldsymbol{a}}) = sgn(\sigma) \cdot (\sigma^{-1})^*(v_{\boldsymbol{a}}) = sgn(\sigma) \cdot sgn(\sigma^{-1}) \cdot v_{\sigma \boldsymbol{a}} = v_{\boldsymbol{b}}.$$

This proves that φ is surjective, because the v_b form a basis of $V_{[\boldsymbol{a}]}$. But then φ must be injective as well, because source and target of φ have the same dimension (namely $(S_n : (S_n)\boldsymbol{a}) = \#[\boldsymbol{a}]$), and the proof of the first statement is complete.

The formula for the trace then follows from the following general formula for the character of an induced representation from [Lan93, XVIII,§6,p686]:

Let H be a group, let S be a subgroup of H, let χ be a character of S, and let χ^H be the induced character of H. Then for all $h \in H$

$$\chi^H(h) = \frac{1}{\#S} \sum_{g \in H} \chi(ghg^{-1}) \quad \text{(with } \chi(t) := 0 \text{ for } t \notin S\text{)}.$$

For $H := S_n$, $S := (S_n)\boldsymbol{a}$, $g := \mathbf{sgn}$ and $h := \sigma$ we get

$$\text{Tr}\left(\sigma^* \mid V_{[\boldsymbol{a}]}\right) \overset{(11.17)}{=} \mathbf{sgn}^{S_n}(\sigma)$$

$$= \frac{1}{\#(S_n)\boldsymbol{a}} \sum_{\tau \in S_n} \underbrace{\mathbf{sgn}(\tau\sigma\tau^{-1})}_{= \begin{cases} sgn(\sigma) \text{ if } \tau\sigma\tau^{-1} \in (S_n)\boldsymbol{a} \\ 0 \qquad \text{otherwise} \end{cases}}$$

$$= \mathbf{sgn}(\sigma) \cdot \frac{\#\left\{\tau \in S_n \,\middle|\, \tau\sigma\tau^{-1} \in (S_n)\boldsymbol{a}\right\}}{\#(S_n)\boldsymbol{a}}.$$

To prove the last equation, we use the isomorphism of S_n-sets

$$S_n/(S_n)\boldsymbol{a} \overset{\sim}{\to} [\boldsymbol{a}], \quad \tau \cdot (S_n)\boldsymbol{a} \mapsto {}^\tau \boldsymbol{a}$$

and get

$$\left\{\boldsymbol{b} \in [\boldsymbol{a}] \,\middle|\, {}^\sigma\boldsymbol{b} = \boldsymbol{b}\right\} \cong \left\{\tau \cdot (S_n)\boldsymbol{a} \in S_n/(S_n)\boldsymbol{a} \,\middle|\, \sigma\tau \cdot (S_n)\boldsymbol{a} = \tau \cdot (S_n)\boldsymbol{a}\right\}$$

$$= \left\{\tau \cdot (S_n)\boldsymbol{a} \in S_n/(S_n)\boldsymbol{a} \,\middle|\, \tau^{-1}\sigma\tau \in (S_n)\boldsymbol{a}\right\}.$$

Counting elements finishes the proof. $\qquad\qquad\qquad\qquad\qquad\square$

Corollary 11.5 *Let* $s := (\zeta_1, \ldots, \zeta_n) \cdot \sigma \in \tilde{A}$ *and* $a \in \check{A}$ *be arbitrary elements.*

If $a \in \check{A} \setminus A_n^r$, *then* s^* *acts trivially on* $V_{[a]}$. *If* $a \in A_n^r$ *and if* $\{v_b\}_{b \in [a]}$ *is a basis of* $V_{[a]}$ *as in corollary 11.3, then for all* $b \in [a]$, *we have:*

$$\boxed{s^* v_b = (\zeta_1^{b_1} \cdot \ldots \cdot \zeta_n^{b_n}) \cdot \text{sgn}(\sigma) \cdot v_{\sigma^{-1} b}}$$

In particular we see that $s^*|V_{[a]}$ *lies in the subgroup*

$$(\pm \mu_r) \wr S([a]) \hookrightarrow \text{Aut}_{\mathbb{Q}_l}(V_{[a]}),$$

where $(\pm \mu_r)$ *denotes the subgroup* $\{\pm \xi \mid \xi^r = 1\}$ *of* $\mathbb{Q}_l^\times$, *where* $S([a])$ *denotes the group of permutations of the set* $[a]$ *and where the injection into* $\text{Aut}_{\mathbb{Q}_l}(V_{[a]})$ *is the one defined in lemma 7.1*

Proof. This follows immediately from example 10.4(2), lemma 11.6(2) and corollary 11.3 (note that $s^* = \sigma^* \circ (\zeta_i)^*$). $\qquad\square$

Example 11.3 Consider the special case $r = 3$, $n = 6$, $s = (\zeta, 1, 1, 1, \zeta^2, 1) \cdot [1234][56]$ (for a primitive third root of unity ζ) and $a = [1, 1, 1, 2, 2, 2]$. Then

$$[a] = \Big\{ \underbrace{[1,1,1,2,2,2]}_{=a=:b_1}, \underbrace{[1,1,2,1,2,2]}_{=:b_2}, \underbrace{[1,1,2,2,1,2]}_{=:b_3}, \underbrace{[1,1,2,2,2,1]}_{=:b_4},$$

$$\underbrace{[1,2,1,1,2,2]}_{=:b_5}, \underbrace{[1,2,1,2,1,2]}_{=:b_6}, \underbrace{[1,2,1,2,2,1]}_{=:b_7}, \underbrace{[1,2,2,1,1,2]}_{=:b_8},$$

$$\underbrace{[1,2,2,1,2,1]}_{=:b_9}, \underbrace{[1,2,2,2,1,1]}_{=:b_{10}}, \underbrace{[2,1,1,1,2,2]}_{=:b_{11}}, \underbrace{[2,1,1,2,1,2]}_{=:b_{12}},$$

$$\underbrace{[2,1,1,2,2,1]}_{=:b_{13}}, \underbrace{[2,1,2,1,1,2]}_{=:b_{14}}, \underbrace{[2,1,2,1,2,1]}_{=:b_{15}}, \underbrace{[2,1,2,2,1,1]}_{=:b_{16}},$$

$$\underbrace{[2,2,1,1,1,2]}_{=:b_{17}}, \underbrace{[2,2,1,1,2,1]}_{=:b_{18}}, \underbrace{[2,2,1,2,1,1]}_{=:b_{19}}, \underbrace{[2,2,2,1,1,1]}_{=:b_{20}} \Big\}.$$

Let $\{v_1, \ldots, v_{20}\}$ be a basis of $V_{[a]}$ as in corollary 11.3 with $v_i \in V_{b_i}$. With respect to this basis, the automorphism $s^*|V_{[a]}$ is then given by the following matrix:

$$
\begin{pmatrix}
0 & 0 & 0 & 0 & 0 & 0 & 0 & 0 & 0 & 0 & \zeta^2 & 0 & 0 & 0 & 0 & 0 & 0 & 0 & 0 & 0 \\
\zeta^2 & 0 & 0 & 0 & 0 & 0 & 0 & 0 & 0 & 0 & 0 & 0 & 0 & 0 & 0 & 0 & 0 & 0 & 0 & 0 \\
0 & 0 & 0 & 0 & 0 & 0 & 0 & 0 & 0 & 0 & 0 & 0 & 1 & 0 & 0 & 0 & 0 & 0 & 0 & 0 \\
0 & 0 & 0 & 0 & 0 & 0 & 0 & 0 & 0 & 0 & 0 & \zeta^2 & 0 & 0 & 0 & 0 & 0 & 0 & 0 & 0 \\
0 & \zeta^2 & 0 & 0 & 0 & 0 & 0 & 0 & 0 & 0 & 0 & 0 & 0 & 0 & 0 & 0 & 0 & 0 & 0 & 0 \\
0 & 0 & 0 & 0 & 0 & 0 & 0 & 0 & 0 & 0 & 0 & 0 & 0 & 0 & 1 & 0 & 0 & 0 & 0 & 0 \\
0 & 0 & 0 & 0 & 0 & 0 & 0 & 0 & 0 & 0 & 0 & 0 & 0 & \zeta^2 & 0 & 0 & 0 & 0 & 0 & 0 \\
0 & 0 & 0 & 1 & 0 & 0 & 0 & 0 & 0 & 0 & 0 & 0 & 0 & 0 & 0 & 0 & 0 & 0 & 0 & 0 \\
0 & 0 & \zeta^2 & 0 & 0 & 0 & 0 & 0 & 0 & 0 & 0 & 0 & 0 & 0 & 0 & 0 & 0 & 0 & 0 & 0 \\
0 & 0 & 0 & 0 & 0 & 0 & 0 & 0 & 0 & 0 & 0 & 0 & 0 & 0 & 0 & 1 & 0 & 0 & 0 & 0 \\
0 & 0 & 0 & 0 & 1 & 0 & 0 & 0 & 0 & 0 & 0 & 0 & 0 & 0 & 0 & 0 & 0 & 0 & 0 & 0 \\
0 & 0 & 0 & 0 & 0 & 0 & 0 & 0 & 0 & 0 & 0 & 0 & 0 & 0 & 0 & 0 & 0 & \zeta & 0 & 0 \\
0 & 0 & 0 & 0 & 0 & 0 & 0 & 0 & 0 & 0 & 0 & 0 & 0 & 0 & 0 & 0 & 1 & 0 & 0 & 0 \\
0 & 0 & 0 & 0 & 0 & 0 & \zeta & 0 & 0 & 0 & 0 & 0 & 0 & 0 & 0 & 0 & 0 & 0 & 0 & 0 \\
0 & 0 & 0 & 0 & 0 & 1 & 0 & 0 & 0 & 0 & 0 & 0 & 0 & 0 & 0 & 0 & 0 & 0 & 0 & 0 \\
0 & 0 & 0 & 0 & 0 & 0 & 0 & 0 & 0 & 0 & 0 & 0 & 0 & 0 & 0 & 0 & 0 & 0 & \zeta & 0 \\
0 & 0 & 0 & 0 & 0 & 0 & 0 & 0 & \zeta & 0 & 0 & 0 & 0 & 0 & 0 & 0 & 0 & 0 & 0 & 0 \\
0 & 0 & 0 & 0 & 0 & 0 & 0 & 1 & 0 & 0 & 0 & 0 & 0 & 0 & 0 & 0 & 0 & 0 & 0 & 0 \\
0 & 0 & 0 & 0 & 0 & 0 & 0 & 0 & 0 & 0 & 0 & 0 & 0 & 0 & 0 & 0 & 0 & 0 & 0 & \zeta \\
0 & 0 & 0 & 0 & 0 & 0 & 0 & 0 & 0 & \zeta & 0 & 0 & 0 & 0 & 0 & 0 & 0 & 0 & 0 & 0
\end{pmatrix}
$$

which, considered as an element of $(\pm\mu_r)^{[a]} \rtimes S([a]))$, equals

$$(\zeta^2, \zeta^2, 1, \zeta^2, \zeta^2, 1, \zeta^2, 1, \zeta^2, 1, 1, \zeta, 1, \zeta, 1, \zeta, \zeta, 1, \zeta, \zeta)$$
$$\cdot \, [1, 2, 5, 11][3, 9, 17, 13][4, 8, 18, 12][6, 15][7, 14][10, 20, 19, 16].$$

(Note that $sgn([1234][56]) = 1$.)

If k contains the r-th roots of unity, we can use equation (6.3), proposition 11.3 and corollary 11.5 to explicitly compute the Frobenius action on the cohomology of an arbitrary twisted Fermat hypersurface over k (we will see an example for that in the next chapter).

If on the other hand k does *not* contain the r-th roots of unity, we unfortunately face a problem: Even though we know how both the Frobenius and elements of $\check{A}$ act with respect to the bases from proposition 11.4 respectively corollary 11.3, this only gives information on how their composition acts if we can choose the two types of bases in a compatible way.

It is easy to see that this is possible if and only if there are no $\boldsymbol{a} \in A_n^r$, $\sigma \in S_n \setminus \{\text{id}\}$ and $t \in (Z/rZ)^\times \setminus \{1\}$ with $t\boldsymbol{a} = {}^\sigma\boldsymbol{a}$.

Such a "good" case is for example the case $r = n = 3$, since $A_3^3 = \{(1,1,1),(2,2,2)\}$, on which S_3 acts trivially and $(\mathbb{Z}/3\mathbb{Z})^\times$ by swapping the two elements.

In contrast to that, the cases $r = 3$, $n = 2$ and $r = 3$, $n = 4$ are "bad": We have $2 \cdot [1,2] = {}^{[12]}[1,2]$ in the first case and $2 \cdot [1,1,2,2] = {}^{[13][24]}[1,1,2,2]$ in the second. Even though we do not yet know how to handle these "bad" cases in general, there are tricks to circumvent the problem in a couple of cases like the two examples above (and thus in particular in the case of binary cubic equations), but we will not pursue this topic any further in this book.

Lemma 11.8 *Let R be a commutative ring with unit, let $N \in \mathbb{N}_+$ be a natural number, let $a_1, \ldots, a_N \in R^\times$ be arbitrary units in R, and let A be the matrix*

$$
\left(
\begin{array}{cccc|c}
0 & 0 & \cdots & 0 & a_N \\
a_1 & 0 & \cdots & 0 & 0 \\
0 & a_2 & \cdots & 0 & 0 \\
\vdots & \vdots & \ddots & \vdots & \vdots \\
0 & 0 & \cdots & a_{N-1} & 0
\end{array}
\right)
$$

over R. Then

$$
\det\left(1 - AT\right) = 1 - \left(\prod_{i=1}^{N} a_i\right) T^N.
$$

In particular, if R is a field, then the eigenvalues of A (in an algebraic closure of R) are (with ζ_N a primitive N-th root of unity)

$$
\left\{ \sqrt[N]{\prod_{i=1}^{N} a_i} \cdot \zeta_N^j \right\}_{j=0,\ldots,N-1}.
$$

Proof. Expansion along the first row gives:

$$\det (1 - AT) = \det \left(\begin{array}{cccc|c} 1 & 0 & \cdots & 0 & -a_N t \\ \hline -a_1 T & 1 & \cdots & 0 & 0 \\ 0 & -a_2 T & \ddots & 0 & 0 \\ \vdots & \vdots & \ddots & \ddots & \vdots \\ 0 & 0 & \cdots & -a_{N-1}T & 1 \end{array}\right)$$

$$= \left|\begin{array}{ccccc} 1 & 0 & \cdots & 0 & 0 \\ -a_2 T & 1 & \cdots & 0 & 0 \\ \vdots & \ddots & \ddots & \vdots & \vdots \\ 0 & 0 & \ddots & 1 & 0 \\ 0 & 0 & \cdots & -a_{N-1}T & 1 \end{array}\right| + (-1)^N a_N T \cdot \left|\begin{array}{ccccc} -a_1 T & 1 & \cdots & 0 & 0 \\ 0 & -a_2 T & \ddots & 0 & 0 \\ \vdots & \vdots & \ddots & \ddots & \vdots \\ 0 & 0 & \cdots & -a_{N-2}T & 1 \\ 0 & 0 & \cdots & 0 & -a_{N-1}T \end{array}\right|$$

$$= 1 + (-1)^N a_N T \cdot (-1)^{N-1} a_1 \cdot \ldots a_{N-1} \cdot T^{N-1}$$

$$= 1 + \underbrace{(-1)^{2N-1}}_{=(-1)} a_1 \cdot \ldots \cdot a_N \cdot T^N.$$

$\qquad\qquad\qquad\qquad\qquad\qquad\qquad\qquad\qquad\qquad\qquad\qquad\qquad\qquad \square$

Corollary 11.6 *Let $s = (\zeta^{e_1}, \ldots, \zeta^{e_n}) \cdot \sigma$ be an element of $\tilde{A}$, and let $\boldsymbol{c}$ be an arbitrary element of $\tilde{A}$, let $[[\boldsymbol{c}]] := \{\boldsymbol{c}, {}^\sigma\boldsymbol{c}, {}^{\sigma^2}\boldsymbol{c}, \ldots\} \subseteq [\boldsymbol{c}]$ be the orbit of $\boldsymbol{c}$ under $\langle\sigma\rangle$, and let $V_{[[\boldsymbol{c}]]} := \oplus_{\boldsymbol{b}\in[[\boldsymbol{c}]]} V_{\boldsymbol{b}}$. Then $s^* \circ F^*$ restricts to an endomorphism of $V_{[[\boldsymbol{c}]]}$ which we denote by φ, and with $N := \#[[\boldsymbol{c}]]$ and $d_{\boldsymbol{c}}^s := \sum_{j=1}^n e_j \sum_{j=0}^{N-1} c_{\sigma^j(i)} \pmod{r}$ we have*

$$det (1 - \varphi T) = 1 - \left(\mathcal{J}_k^r(\boldsymbol{c})^N \cdot sgn(\sigma)^N \cdot \zeta^{d_{\boldsymbol{c}}^s}\right) T^N.$$

Consequently, the eigenvalues of φ are

$$\left\{\mathcal{J}_k^r(\boldsymbol{c}) \cdot sgn(\sigma) \cdot \zeta_{rN}^{d_{\boldsymbol{c}}^s + rj}\right\}_{j=0,\ldots,N-1},$$

where ζ_{rN} is a primitive (rN)-th root of unity.

Proof. This is obvious from corollary 11.5 and lemma 11.8 (note that F^* is just multiplication by $\mathcal{J}_k^r(\boldsymbol{a})$ on $V_{[[\boldsymbol{a}]]}$).

$\qquad\qquad\qquad\qquad\qquad\qquad\qquad\qquad\qquad\qquad\qquad\qquad\qquad\qquad \square$

Example 11.4   In the situation of examples 11.1 and 11.3, we find

c	$[[c]]$	N	$d_{\boldsymbol{c}}^s$			
b	$\{b\}$	1	$1 \cdot 1 + 2 \cdot 1$	$=$	3	$\equiv 0$
b_1	$\{b_1, b_2, b_5, b_{11}\}$	4	$1 \cdot 5 + 2 \cdot 8$	$=$	21	$\equiv 0$
b_3	$\{b_3, b_9, b_{13}, b_{17}\}$	4	$1 \cdot 6 + 2 \cdot 6$	$=$	18	$\equiv 0$
b_4	$\{b_4, b_8, b_{12}, b_{18}\}$	4	$1 \cdot 6 + 2 \cdot 6$	$=$	18	$\equiv 0$
b_6	$\{b_6, b_{15}\}$	2	$1 \cdot 3 + 2 \cdot 3$	$=$	9	$\equiv 0$
b_7	$\{b_7, b_{14}\}$	2	$1 \cdot 3 + 2 \cdot 3$	$=$	9	$\equiv 0$
b_{10}	$\{b_{10}, b_{16}, b_{19}, b_{20}\}$	4	$1 \cdot 7 + 2 \cdot 4$	$=$	15	$\equiv 0$
$2b$	$\{2b\}$	1	$1 \cdot 2 + 2 \cdot 2$	$=$	6	$\equiv 0$

From example 11.1 we know $\mathcal{J}_k^3(b) = -56 - 21\zeta$, $\mathcal{J}_k^3(2b) = -35 + 21\zeta$ and $\mathcal{J}_k^3(a) = 49$, so we get the following characteristic polynomials and eigenvalues on V:

c	$\det(1 - \varphi T)$	eigenvalues on $V_{[[\boldsymbol{c}]]}$
$[0,0,0,0,0,0]$	$1 - 49T$	$\{49\}$
b	$1 - (-56 - 21\zeta)T$	$\{-56 - 21\zeta\}$
b_1	$1 - 49^4 T^4$	$\{49, -49, 49\zeta_4, -49\zeta_4\}$
b_3	$1 - 49^4 T^4$	$\{49, -49, 49\zeta_4, -49\zeta_4\}$
b_4	$1 - 49^4 T^4$	$\{49, -49, 49\zeta_4, -49\zeta_4\}$
b_6	$1 - 49^2 T^2$	$\{49, -49\}$
b_7	$1 - 49^2 T^2$	$\{49, -49\}$
b_{10}	$1 - 49^4 T^4$	$\{49, -49, 49\zeta_4, -49\zeta_4\}$
$2b$	$1 - (-35 + 21\zeta)T$	$\{-35 + 21\zeta\}$

Thus we have eigenvalues 49 (with multiplicity 7), (-49) (with multiplicity 6), $49\zeta_4$ (with multiplicity 4), $(-49\zeta_4)$ (with multiplicity 4) and $(-56-21\zeta)$ and $(-35 + 21\zeta)$ (with multiplicity one), and we know

$$
\begin{aligned}
\det(1 - s^* F^* T | V) &= (1 - 49T)^7 (1 + 49T)^6 (1 - 49\zeta_4 T)^4 (1 - 49\bar\zeta_4 T)^4 \\
&\quad \cdot (1 - [-56 - 21\zeta]T)(1 - [-35 + 21\zeta]T) \\
&= (1 - 91T + 2401T^2)(1 - 49T)(1 - 2401T^2)^2 (1 - 5764801T^4)^4.
\end{aligned}
$$

Chapter 12

The zeta function of forms of Fermat equations

Finally we are now in a position to compute the zeta function of arbitrary twisted Fermat equations over finite fields (that contain suitable roots of unity). All we have to do is collect our results from the previous chapters and put them together.

After we have explained how to do this, we will give an explicit example.

Let $n, r \in \mathbb{N}_+$ with $n \geq 2$, let $\tilde{A} := \mu_r^n / \mu_r \rtimes S_n$, and let $k = \mathbb{F}_q$ be a finite field of characteristic $p > \max(r, 2)$ with separable closure K and absolute Galois group G — here we want to assume the k contains the r-th roots of unity.

Let Q be an arbitrary twisted Fermat equation of degree r in n variables, i.e. an object of $E(\widehat{\mathcal{F}_K^{n,r}} / \widehat{\mathcal{F}_k^{n,r}}, P_n^r)$. From corollary 9.3 we know that $Q \cong P_n^r\{(L, x)\}$ for a suitable fcs-algebra L of degree n over k and an element $x \in L^\times$.

The class $(L, x) \in H^1_{\text{cont}}(G, \mu_r \wr S_n)$ is represented by a cocycle $b : G \to \tilde{A}$, and using proposition 7.2 or corollary 7.5, we can explicitly compute the value $b_f \in \tilde{A}$ of b at the arithmetic Frobenius $f \in G$.

Now, as in corollary 6.2, we denote the hypersurfaces associated to P_n^r and Q by X respectively Y and identify the $\mathbb{Q}_l$-vector space $H^{n-2}_{\text{ét}}(X \times_k K, \mathbb{Q}_l)$ with $V := H^{n-2}_{\text{ét}}(Y \times_k K, \mathbb{Q}_l)$. Using equation (6.3), we know that the geometrical Frobenii F_X^* and F_Y^* of X respectively Y are related by $F_Y^* = \left(f^{-1}b_f \right)^* \circ F_X^*$. Since we assume that k contains the r-th roots of unity, the arithmetic Frobenius acts trivially on $\tilde{A}$, so the equation simplifies to give $\boxed{F_Y^* = b_f^* \circ F_X^*}$.

The next step is to explicitly compute b_f^* and F_X^* with respect to a suitable basis of V: We do this by applying proposition 11.3 and corollary 11.5 or, if we are only interested in the eigenvalues λ_j or the characteristic polynomial $\det(1 - b_f^* F_X^* T)$, we can use corollary 11.6.

Finally, to compute the zeta function of Q (which is by definition the zeta function of Y), we compute the characteristic polynomial of $b_f^* \circ F_X^*$ on V with corollary 11.6 and use proposition 5.3; if we are interested in the number of $\mathbb{F}_{q^i}$-rational points of Y for an $i \geq 1$, we just have to compute the trace of $(b_f^* \circ F_X^*)^i$ (which is easy since we know the eigenvalues λ_j, so that the trace in question is just $\sum_j \lambda_j^i$) and apply equation (5.13).

Let us see a practical example:

Example 12.1   Let $r = 3$, $n = 6$ and $k = \mathbb{F}_7$. Then $k^\times$ has order six and thus contains the third roots of unity.

Let b be the class $\left(\mathbb{F}_{2401} \times \mathbb{F}_{49}, (\frac{1}{\beta}, \frac{1}{\alpha^2})\right)$ in $H^1_{\mathrm{cont}}(G, \mu_3 \wr S_6)$ from example 7.1. In that example, we have seen that we can take $b_f = (\zeta, 1, 1, 1, \zeta^2, 1) \cdot [1234][56] \in \mu_3 \wr S_6$ (with $\zeta := \zeta_3 := 2 \in \mu_3 \subseteq \mathbb{F}_7^\times$), and in example 7.3, we calculated the K/k-form $Q = P_6^3\{b\}$ of P_6^3 given by b explicitly:

$$
\begin{aligned}
Q = {}& 2a^3 + 6a^2c + a^2d + 6ab^2 + 2abc + 4abd + 2ac^2 + 3acd + 6ad^2 \\
&+ 5b^3 + 2b^2c + 5b^2d + 5bc^2 + 5bcd + 6bd^2 \\
&+ 2c^3 + 6c^2d + cd^2 + d^3 \\
&+ y^3 + 4x^2y + 5y^3.
\end{aligned}
$$

Let Y be the four-dimensional hypersurface in $\mathbb{P}_k^5$ associated to Q, and let $V := H^4_{\text{ét}}(X_6^3, \mathbb{Q}_l)$ be the middle l-adic cohomology. From example 11.1, we know $V = V_{[0,0,0,0,0,0]} \oplus V_{[a]} \oplus V_{b} \oplus V_{2b}$ with $a = [1,1,1,2,2,2]$ and $b = [1,1,1,1,1,1]$ and that $\mathcal{J}_k^3(a) = 49$, $\mathcal{J}_k^3(b) = -56 - 21\zeta$ and $\mathcal{J}_k^3(2b) = -35 + 21\zeta$. In example 11.3, we have computed b_f^* on $V_{[a]}$ explicitly: With respect to a suitable basis of $V_{[a]}$,

$$
\begin{aligned}
b_f^* = {}& (\zeta^2, \zeta^2, 1, \zeta^2, \zeta^2, 1, \zeta^2, 1, \zeta^2, 1, 1, \zeta, 1, \zeta, 1, \zeta, \zeta, 1, \zeta, \zeta) \\
& \cdot [1,2,5,11][3,9,17,13][4,8,18,12][6,15][7,14][10,20,19,16].
\end{aligned}
$$

Furthermore, we know from lemma 11.6 that on $V_{[0,0,0,0,0,0]}$, b_f^* is the identity and F_X^* is multiplication by 49, and finally we know that b_f^* is the identity on V_{b} and V_{2b} because $b([\zeta, 1, 1, 1, \zeta^2, 1]) = 1$.

From the explicit knowledge of F_Y^*, we can compute the characteristic polynomial and the eigenvalues — we have done so in example 11.4 and got

$$\det\left(1 - F_Y^*|V\right)$$
$$= (1 - 91t + 2401t^2)(1 - 49t)(1 - 2401t^2)^2(1 - 5764801t^4)^4,$$

eigenvalue	multiplicity
49	7
−49	6
$49\zeta_4$	4
$-49\zeta_4$	4
$-56 - 21\zeta_3$	1
$-35 + 21\zeta_3$	1

Using proposition 5.3 and the characteristic polynomial of F_Y^* on V, we can eventually compute the zeta function of Q:

$$\zeta(P, T) = \frac{1}{1 - T} \cdot \frac{1}{1 - 7T}$$
$$\cdot \frac{1}{(1 - 91T + 2401T^2)(1 - 49T)(1 - 2401T^2)^2(1 - 5764801T^4)^4}$$
$$\cdot \frac{1}{1 - 7^3T} \cdot \frac{1}{1 - 7^4T}.$$

Finally, we want to calculate the number $\nu_Y^{(i)}$ of non-trivial solutions of Q (i.e. of points of Y) over the fields $\mathbb{F}_{7^i}$ for $i = 1, 2, 3, 4$. For that, we need the traces $t_i := \operatorname{Tr}\left[(F_Y^*)^i|V\right]$, because then according to equation (5.13):

$$\nu_Y^{(i)} = 7^0 + 7^i + 7^{3i} + 7^{4i} + t_i = \underbrace{1 + 7^i + 343^i + 2401^i}_{=:s_i} + t_i.$$

To get the traces, we use the eigenvalues we already know (in the table, m_j denotes the multiplicity of eigenvalue λ_j):

j	m_j	λ_j	λ_j^2	λ_j^3	λ_j^4
1	7	49	2401	117649	5764801
2	6	-49	2401	-117649	5764801
3	4	$49\zeta_4$	-2401	$-117649\zeta_4$	5764801
4	4	$-49\zeta_4$	-2401	$117649\zeta_4$	5764801
5	1	$-56 - 21\zeta_3$	2695 $+1911\zeta_3$	-110789 $-123480\zeta_3$	3611104 $+6648369\zeta_3$
6	1	$-35 + 21\zeta_3$	784 $-1911\zeta_3$	12691 $+123480\zeta_3$	-3037265 $-6648369\zeta_3$
t_i		-42	15484	19551	121634660
s_i		2752	5882500	13881641152	33246771859204
$\nu_Y^{(i)}$			2710	5897984	13881660703	33246893493864

So Y has 2710 points over $\mathbb{F}_7$, 5897984 points over $\mathbb{F}_{49}$, 13881660703 points over $\mathbb{F}_{343}$ and 33246893493864 points over $\mathbb{F}_{2401}$, so the zeta function of Q can also be written as

$$\exp\left(\frac{2710}{1}T + \frac{5897984}{2}T^2 + \frac{13881660703}{3}T^3 + \frac{33246893493864}{4}T^4 + \ldots\right).$$

Even though these calculations may seem somewhat lengthy, just think how long it would take even a fast computer to count these $3.32 \cdot 10^{13}$ points over $\mathbb{F}_{2401}$ by brute force!

Appendix A

Pseudo-abelian categories

Let R be a commutative ring with unit.

Definition A.1   An R-*linear category* is a category $\mathcal{M}$ with zero object and finite sums which is endowed with an R-module structure on all morphism sets $\mathrm{Mor}_{\mathcal{M}}(X, Y)$ of $\mathcal{M}$ in such a way that all compositions $\mathrm{Mor}_{\mathcal{M}}(Y, Z) \times \mathrm{Mor}_{\mathcal{M}}(X, Y) \to \mathrm{Mor}_{\mathcal{M}}(X, Z)$ become R-bilinear.

(Note that according to this definition, a $\mathbb{Z}$-linear category is just an additive category in the usual sense.)

Example A.2   The category of R-modules is an R-linear category if we set $(f + g)(x) := f(x) + g(x)$ and $(rf)(x) := rf(x)$ for R-modules M and N, homomorphisms $f, g \in \mathrm{Hom}_R(M, N)$, a scalar $r \in R$ and an element $x \in M$.

Lemma A.1   *Let $\mathcal{M}$ be an R-linear category, let $\beta : X \to Y$ be an arbitrary morphism in $\mathcal{M}$, let $\alpha : Y \hookrightarrow Z$ be a monomorphism in $\mathcal{M}$, and assume that the kernel $\bar{\alpha} : \mathrm{Ker}\,(\alpha \circ \beta) \hookrightarrow X$ of $(\alpha \circ \beta)$ exists in $\mathcal{M}$. Then $\bar{\alpha}$ is also the kernel of β.*

Proof.   Put $\bar{p} := \alpha \circ \beta$. We have

$$0 = \bar{p} \circ \bar{\alpha} = \alpha \circ \left[\beta \circ \bar{\alpha}\right],$$

so $\beta \circ \bar{\alpha} = 0$, since α is a monomorphism. Now let $f : W \to X$ be a morphism in $\mathcal{M}$ with $\beta \circ f = 0$. We have to show that there is a unique morphism $\bar{f} : W \to \mathrm{Ker}\,(\bar{p})$ which makes the following diagram in $\mathcal{M}$ commute:

$$W \xdashrightarrow{\bar{f}} \mathrm{Ker}\,(\bar{p})$$
$$\searrow_{f} \qquad \downarrow_{\bar{\alpha}}$$
$$X$$

(A.1)

But because of $\beta \circ f = 0$, we also have $\bar{p} \circ f = \alpha \circ (\beta \circ f) = 0$, so the universal property of $\bar{\alpha}$ implies that there is indeed a unique morphism $\bar{f}$ making (A.1) commute. □

Definition A.2 Let $\mathcal{C}$ be a category, and let X be an object of $\mathcal{C}$. An endomorphism $p \in \mathrm{End}_{\mathcal{C}}(X)$ of X is called a *projector* if $p \circ p = p$, i.e. if p is idempotent.

Proposition A.1 *Let $\mathcal{M}$ be an R-linear category, let X be an object of X, and let $p \in \mathrm{End}_{\mathcal{M}}(X)$ be a projector. Then*

(1) The endomorphism $\bar{p} := (1_X - p)$ of X is also a projector.

(2) If the kernel of p exists, then $\bar{p}$ factorizes over the canonical embedding $\alpha : \mathrm{Ker}(p) \hookrightarrow X$, and the induced morphism $\beta : X \to \mathrm{Ker}(p)$ is a section of α and the cokernel of p.

(3) If both the kernels of p and of $\bar{p}$ exist, if $\bar{\alpha} : \mathrm{Ker}(\bar{p}) \hookrightarrow X$ denotes the canonical embedding, and if $\bar{\beta} : X \to \mathrm{Ker}(\bar{p})$ is the morphism from (2) for $\bar{p}$ instead of p, then

    (a) $\bar{\alpha}$ is the image of p,

    (b) we have a canonical factorization

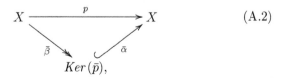

$$X \xrightarrow{\quad p \quad} X$$
$$\searrow_{\bar{\beta}} \qquad \nearrow_{\bar{\alpha}}$$
$$\mathrm{Ker}\,(\bar{p}),$$

(A.2)

    (c) the canonical morphism

$$[\mathrm{Ker}\,(p)] \oplus [\mathrm{Ker}\,(\bar{p})] \xrightarrow{\alpha \oplus \bar{\alpha}} X$$

    is an isomorphism with inverse $(\beta, \bar{\beta})$,

(d) and we have (if we identify $Im\,(p)$ with $Ker\,(\bar{p})$ using (a))

$$Ker\,(p) \overset{\alpha}{\hookrightarrow} X \overset{p}{\to} X \overset{\beta}{\to} Ker\,(p) = 0,$$
$$Ker\,(p) \overset{\alpha}{\hookrightarrow} X \overset{p}{\to} X \overset{\bar{\beta}}{\to} Im\,(p) = 0,$$
$$Im\,(p) \overset{\bar{\alpha}}{\hookrightarrow} X \overset{p}{\to} X \overset{\beta}{\to} Ker\,(p) = 0,$$
$$Im\,(p) \overset{\bar{\alpha}}{\hookrightarrow} X \overset{p}{\to} X \overset{\bar{\beta}}{\to} Im\,(p) = 1_{Im\,(p)}.$$

Proof. First of all, because composition in $\mathcal{M}$ is R-bilinear, we have

$$\bar{p} \circ \bar{p} = (1_X - p) \circ (1_X - p)$$
$$= (1_X \circ 1_X) - (1_X \circ p) - (p \circ 1_X) + (p \circ p)$$
$$= 1_X - p - p + p = 1_X - p = \bar{p},$$

so $\bar{p}$ is indeed a projector and (1) is proved. Now assume that the kernel of p exists. We have

$$p \circ \bar{p} = p \circ (1_X - p) = (p \circ 1_X) - (p \circ p) = p - p = 0,$$

so $\bar{p}$ uniquely factorizes over $\alpha : Ker\,(p) \hookrightarrow X$. To show that the induced morphism $\beta : X \to Ker\,(p)$ is a section of α, we calculate:

$$\alpha \circ (\beta \circ \alpha) = (\alpha \circ \beta) \circ \alpha = \bar{p} \circ \alpha = (1_X - p) \circ \alpha$$
$$= (1_X \circ \alpha) - (p \circ \alpha) = \alpha - 0 = \alpha = \alpha \circ 1_{Ker\,(p)},$$

so because α is a monomorphism, it follows that $\beta \circ \alpha = 1_{Ker\,(p)}$, which means that β is really a section of α. Now we want to show that β is the cokernel of p:

$$\alpha \circ (\beta \circ p) = (\alpha \circ \beta) \circ p = \bar{p} \circ p$$
$$= (1_X - p) \circ p = (1_X \circ p) - (p \circ p) = p - p = 0,$$

so $\beta \circ p = 0$, because α is a monomorphism. Let $f : X \to Y$ be a morphism in $\mathcal{M}$ with $f \circ p = 0$. We must show that there is a unique morphism $\bar{f} : Ker\,(p) \to Y$ such that the following diagram commutes in $\mathcal{M}$:

$$X \overset{\beta}{\longrightarrow} Ker\,(p) \qquad (\text{A.3})$$

with arrows f from X to Y and $\bar{f}$ from $Ker\,(p)$ to Y.

Put $\bar{f} := f \circ \alpha$, then we get

$$\bar{f} \circ \beta = (f \circ \alpha) \circ \beta = f \circ (\alpha \circ \beta) = f \circ \bar{p} = f \circ (1_X - p)$$
$$= (f \circ 1_X) - (f \circ p) = f - 0 = f,$$

so diagram (A.3) commutes. To prove uniqueness, let $f' : \mathrm{Ker}\,(p) \to Y$ be another morphism with $f' \circ \beta = f$. Then

$$0 = (\bar{f} \circ \beta) - (f' \circ \beta) = (\bar{f} - f') \circ \beta$$

and thus

$$0 = \left[(\bar{f} - f') \circ \beta \right] \circ \alpha = (\bar{f} - f') \circ [\beta \circ \alpha] = (\bar{f} - f') \circ 1_{\mathrm{Ker}\,(p)} = \bar{f} - f',$$

hence $\bar{f} = f'$. This completes the proof of (2). Now assume that the kernel of $\bar{p}$ also exists.

(a) By definition of the image, we have to show that $\bar{\alpha}$ is the kernel of the cokernel of p, where the latter is β because of (2). But this follows immediately from lemma A.1, because α is a monomorphism and because $\bar{p} = \alpha \circ \beta$.

(b) In order to show that diagram (A.2) commutes, we remember from (2) that $\bar{p} = \alpha \circ \beta$, which implies $p = \bar{\bar{p}} = \bar{\alpha} \circ \bar{\beta}$ by replacing p with $\bar{p}$.

(c) We claim that

$$X \xrightarrow{(\beta,\bar{\beta})} [\mathrm{Ker}\,(p)] \oplus [\mathrm{Ker}\,(\bar{p})]$$

is inverse to $\alpha \oplus \bar{\alpha}$. First note that the following diagram commutes:

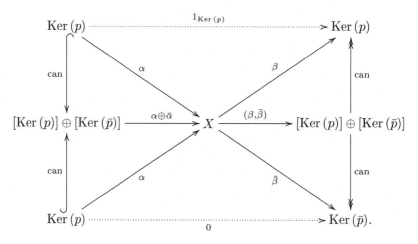

(We have $\beta \circ \alpha = 1_{\mathrm{Ker}\,(p)}$ by (2) and

$$\bar{\alpha} \circ \left[\bar{\beta} \circ \alpha\right] = \left[\bar{\alpha} \circ \bar{\beta}\right] \circ \alpha = p \circ \alpha = 0,$$

so $\bar{\beta} \circ \alpha = 0$, because $\bar{\alpha}$ is a monomorphism.) Because the diagram commutes, we see that

$$\mathrm{Ker}\,(p) \xrightarrow{\mathrm{can}} [\mathrm{Ker}\,(p)] \oplus [\mathrm{Ker}\,(\bar{p})] \xrightarrow{(\beta,\bar{\beta})\circ[\alpha\oplus\bar{\alpha}]} [\mathrm{Ker}\,(p)] \oplus [\mathrm{Ker}\,(\bar{p})]$$

is the morphism $(1_{\mathrm{Ker}\,(p)}, 0)$, and replacing p by $\bar{p}$, we see that

$$\mathrm{Ker}\,(\bar{p}) \xrightarrow{\mathrm{can}} [\mathrm{Ker}\,(p)] \oplus [\mathrm{Ker}\,(\bar{p})] \xrightarrow{(\beta,\bar{\beta})\circ[\alpha\oplus\bar{\alpha}]} [\mathrm{Ker}\,(p)] \oplus [\mathrm{Ker}\,(\bar{p})]$$

is the morphism $(0, 1_{\mathrm{Ker}\,(\bar{p})})$, so it follows that $(\beta, \bar{\beta}) \circ [\alpha \oplus \bar{\alpha}]$ is the identity on $[\mathrm{Ker}\,(p)] \oplus [\mathrm{Ker}\,(\bar{p})]$. Finally, the diagram

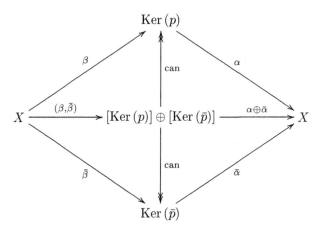

commutes, which shows that

$$[\alpha \oplus \bar{\alpha}] \circ (\beta, \bar{\beta}) = \left[\alpha \circ \beta\right] + \left[\bar{\alpha} \circ \bar{\beta}\right] = \bar{p} + p = (1_X - p) + p = 1_X.$$

This finishes the proof of the fact that $(\beta, \bar{\beta})$ is inverse to $\alpha \oplus \bar{\alpha}$.

(d) The first two equations are obvious, because $p \circ \alpha = 0$ by the definition of "kernel". For the other two equations, we write p as $\bar{\alpha} \circ \bar{\beta}$ according to (b) and get

$$\beta \circ p \circ \bar{\alpha} = \beta \circ \left[\bar{\alpha} \circ \bar{\beta}\right] \circ \bar{\alpha} = \left[\beta \circ \bar{\alpha}\right] \circ \left[\bar{\beta} \circ \bar{\alpha}\right] \overset{(c)}{=} 0 \circ 1_{\mathrm{Im}\,(p)} = 0,$$

$$\bar{\beta} \circ p \circ \bar{\alpha} = \bar{\beta} \circ \left[\bar{\alpha} \circ \bar{\beta}\right] \circ \bar{\alpha} = \left[\bar{\beta} \circ \bar{\alpha}\right] \circ \left[\bar{\beta} \circ \bar{\alpha}\right]$$

$$\overset{(c)}{=} 1_{\mathrm{Im}\,(p)} \circ 1_{\mathrm{Im}\,(p)} = 1_{\mathrm{Im}\,(p)}.$$

So the proof of (3) and hence the proof of the proposition is complete. □

Definition A.3   Let R be a ring. An R-linear category $\mathcal{M}$ is called *pseudo-abelian* if for all objects $X \in \mathrm{Ob}(\mathcal{M})$, all projectors in $\mathrm{End}_{\mathcal{M}}(X)$ have a kernel.

Corollary A.1   *Let $\mathcal{M}$ be a pseudo-abelian R-linear category, let X be an object of $\mathcal{M}$, and let $p \in End_{\mathcal{M}}(X)$ be a projector. Then the image of p exists, the canonical morphism $Ker\,(p) \oplus Im\,(p) \to X$ is an isomorphism, and the following diagram commutes in $\mathcal{M}$:*

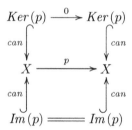

Proof.   Because of proposition A.1(1), $(1_X - p)$ is a projector, too, and since $\mathcal{M}$ is pseudo-abelian, the kernels of both p and $(1_X - p)$ exist. Then the corollary follows immediately from proposition A.1(3).   □

Example A.3   Any *abelian* R-linear category is obviously pseudo-abelian, because in abelian categories, *all* morphisms have a kernel. So the category of R-modules is an example of an R-linear pseudo-abelian category when endowed with canonical structure of an R-linear category as in example A.2.

Proposition A.2   *Let $\mathcal{M}$ be an R-linear pseudo-abelian category, let X be an object of $\mathcal{M}$, and let $\{p_1, \ldots, p_s\}$ be a complete system of idempotents of $End_{\mathcal{M}}(X)$. Then the canonical morphism*

$$\bigoplus_{i=1}^{s} Im\,(p_i) \longrightarrow X$$

is an isomorphism (the images exist because of corollary A.1).

Proof. We prove the proposition by induction over s: For $s = 1$, we have $p_1 = 1_X$ and the statement is trivial, so let $s \in \mathbb{N}_+$ be an arbitrary natural number, let us assume that the lemma holds for s, and let $\{p_1, \ldots, p_s\}$ a complete set of idempotents of $\text{End}_{\mathcal{M}}(X)$. Set $p := \sum_{i=1}^{s} p_i \in \text{End}_{\mathcal{M}}(X)$. Then p is a projector:

$$ p \circ p = \left(\sum_{i=1}^{s} p_i \right) \circ \left(\sum_{i=1}^{s} p_i \right) = \sum_{i,j=1}^{s} \underbrace{p_i \circ p_j}_{= \delta_{ij} p_i} = \sum_{i=1}^{s} p_i = p. $$

Let $\bar{\alpha} : X_p \hookrightarrow X$ be the kernel of the p_{s+1} — then $\bar{\alpha}$ is also the image of p by proposition A.1(3a), because $p_{s+1} = 1_X - p$. For each $i \in \{1, \ldots, s\}$, we have $p_{s+1} \circ p_i = 0$, so the universal property of the kernel implies that p_i uniquely factorizes over a morphism $\beta_i : X \to X_p$, and by restriction we get an endomorphism q_i of X_p:

$$ \begin{array}{ccccc} X_p & \overset{\bar{\alpha}}{\hookrightarrow} & X & \overset{p_i}{\longrightarrow} & X \\ {\scriptstyle q_i} \downarrow & & {\scriptstyle \beta_i} \downarrow & \nearrow {\scriptstyle \bar{\alpha}} & \\ X_p & = & \text{Ker}\,(p_{s+1}) & & \end{array} \qquad (A.4) $$

We claim that $\{q_1, \ldots, q_s\}$ is a complete system of idempotents in $\text{End}_{\mathcal{M}}(X_p)$: For $i, j \in \{1, \ldots, s\}$, the diagrams

$$ \begin{array}{ccccc} X_p & \overset{q_j}{\longrightarrow} & X_p & \overset{q_i}{\longrightarrow} & X_p \\ {\scriptstyle \bar{\alpha}} \uparrow & & {\scriptstyle \bar{\alpha}} \uparrow & & {\scriptstyle \bar{\alpha}} \uparrow \\ X & \underset{p_j}{\longrightarrow} & X & \underset{p_i}{\longrightarrow} & X \\ & \underset{\delta_{ij} p_i}{\searrow\nearrow} & & & \end{array} \qquad \begin{array}{ccc} X_p & \overset{\delta_{ij} q_i}{\longrightarrow} & X_p \\ {\scriptstyle \bar{\alpha}} \uparrow & & {\scriptstyle \bar{\alpha}} \uparrow \\ X & \underset{\delta_{ij} p_i}{\longrightarrow} & X \end{array} $$

commute, so we get $q_i \circ q_j = \delta_{ij} q_i$, since $\bar{\alpha}$ is a monomorphism. Furthermore, the diagrams

$$
\begin{array}{ccc}
X_p & \xrightarrow{\sum_{i=1}^{s} q_i} & X_p \\
\bar{\alpha} \uparrow & & \uparrow \bar{\alpha} \\
X & \xrightarrow{\sum_{i=1}^{s} p_i} & X
\end{array}
\qquad
\begin{array}{ccc}
X_p & = & X_p \\
\bar{\alpha} \uparrow & & \uparrow \bar{\alpha} \\
X & \xrightarrow{\;p\;} & X
\end{array}
$$

commute (the right one by proposition A.1(3d)), so as above it follows that $\sum_{i=1}^{s} q_i = 1_{X_p}$. Thus we have seen that $\{q_1, \dots, q_s\}$ is indeed a complete system of idempotents in $\mathrm{End}_{\mathcal{M}}(X_p)$.

By the inductive hypothesis, applied to X_p and $\{q_1, \dots, q_s\}$, we know that the canonical morphism

$$
\bigoplus_{i=1}^{s} \mathrm{Im}\,(q_i) \longrightarrow X_p \tag{A.5}
$$

is an isomorphism. Now we want to see that for $i \in \{1, \dots, s\}$ the composition $\mathrm{Im}\,(q_i) \overset{\mathrm{can}}{\hookrightarrow} X_p \overset{\bar{\alpha}}{\to} X$ is the image of p_i. By proposition A.1(3a), we know that $\mathrm{Im}\,(q_i) = \mathrm{Ker}\,(1_{X_p} - q_i)$ and $\mathrm{Im}\,(p_i) = \mathrm{Ker}\,(1_X - p_i)$, so what we have to show is that there is an isomorphism $\gamma_i : \mathrm{Ker}\,(1_{X_p} - q_i) \to \mathrm{Ker}\,(1_X - p_i)$ such that the following diagram commutes:

$$
\begin{array}{ccc}
\mathrm{Ker}\,(1_{X_p} - q_i) & \overset{\mathrm{can}}{\hookrightarrow} & X_p \\
\gamma_i \downarrow \wr & & \uparrow \bar{\alpha} \\
\mathrm{Ker}\,(1_X - p_i) & \underset{\mathrm{can}}{\hookrightarrow} & X.
\end{array}
\tag{A.6}
$$

We have

$$
(1_X - p_i) \circ \bar{\alpha} = \bar{\alpha} - \left[p_i \circ \bar{\alpha}\right] \overset{(A.4)}{=} \bar{\alpha} - \left[\left(\bar{\alpha} \circ \beta_i\right) \circ \bar{\alpha}\right]
$$

$$
= \bar{\alpha} - \left[\bar{\alpha} \circ \left(\beta_i \circ \bar{\alpha}\right)\right] \overset{(A.4)}{=} \bar{\alpha} - \left[\bar{\alpha} \circ q_i\right] = \bar{\alpha} \circ \left(1_{X_p} - q_i\right),
$$

so by the universal property of the kernel, we get a morphism γ_i such that the following diagram commutes:

$$
\begin{array}{ccccc}
\mathrm{Ker}\,(1_{X_p} - q_i) & \xrightarrow{\;\mathrm{can}\;} & X_p & \xrightarrow{\;1_{X_p}-q_i\;} & X_p \\
\downarrow{\scriptstyle \gamma_i} & & \downarrow{\scriptstyle \bar{\alpha}} & & \downarrow{\scriptstyle \bar{\alpha}} \\
\mathrm{Ker}\,(1_X - p_i) & \xrightarrow[\;\mathrm{can}\;]{} & X & \xrightarrow[\;1_X-p_i\;]{} & X.
\end{array}
$$

So in particular diagram (A.6) commutes, and we only have to see that γ_i is an isomorphism. Using diagram (A.4) again, we get

$$
(1_{X_p} - q_i) \circ \beta_i = \beta_i - \big[q_i \circ \beta_i\big] \overset{(\mathrm{A.4})}{=} \beta_i - \Big[\big(\beta_i \circ \bar{\alpha}\big) \circ \beta_i\Big]
$$
$$
= \beta_i - \Big[\beta_i \circ \big(\bar{\alpha} \circ \beta_i\big)\Big] = \beta_i - \big[\beta_i \circ p_i\big] = \beta_i \circ (1_X - p_i),
$$

so by the universal property of the kernel, we get a morphism δ_i such that the following diagram commutes:

$$
\begin{array}{ccccc}
\mathrm{Ker}\,(1_X - p_i) & \xrightarrow{\;\mathrm{can}\;} & X & \xrightarrow{\;1_X-p_i\;} & X \\
\downarrow{\scriptstyle \delta_i} & & \downarrow{\scriptstyle \beta_i} & & \downarrow{\scriptstyle \beta_i} \\
\mathrm{Ker}\,(1_{X_p} - q_i) & \xrightarrow[\;\mathrm{can}\;]{} & X_p & \xrightarrow[\;1_{X_p}-q_i\;]{} & X_p.
\end{array}
$$

We claim that δ_i is inverse to γ_i. By diagram (A.4), the following two diagrams commute:

$$
\begin{array}{ccc}
\mathrm{Im}\,(p_i) & \xrightarrow{\;\mathrm{can}\;} & X \\
{\scriptstyle \delta_i}\downarrow & & \downarrow{\scriptstyle \beta_i} \\
\mathrm{Im}\,(q_i) & \xrightarrow{\;\mathrm{can}\;} & X_p \\
{\scriptstyle \gamma_i}\downarrow & & \downarrow{\scriptstyle \bar{\alpha}} \\
\mathrm{Im}\,(p_i) & \xrightarrow[\;\mathrm{can}\;]{} & X
\end{array}
\qquad
\begin{array}{ccc}
\mathrm{Im}\,(q_i) & \xrightarrow{\;\mathrm{can}\;} & X_p \\
{\scriptstyle \gamma_i}\downarrow & & \downarrow{\scriptstyle \bar{\alpha}} \\
\mathrm{Im}\,(p_i) & \xrightarrow{\;\mathrm{can}\;} & X \\
{\scriptstyle \delta_i}\downarrow & & \downarrow{\scriptstyle \beta_i} \\
\mathrm{Im}\,(q_i) & \xrightarrow[\;\mathrm{can}\;]{} & X_p,
\end{array}
$$

$\gamma_i \circ \delta_i$ on the left, $\delta_i \circ \gamma_i$ on the right, with p_i and q_i as the outer arrows,

and by proposition A.1(3d), the following two diagrams commute:

$$
\begin{array}{ccc}
\mathrm{Im}\,(p_i) & \xrightarrow{\;\mathrm{can}\;} & X \\
\Big\| & & \downarrow{\scriptstyle p_i} \\
\mathrm{Im}\,(p_i) & \xrightarrow[\;\mathrm{can}\;]{} & X
\end{array}
\qquad
\begin{array}{ccc}
\mathrm{Im}\,(q_i) & \xrightarrow{\;\mathrm{can}\;} & X_p \\
\Big\| & & \downarrow{\scriptstyle q_i} \\
\mathrm{Im}\,(q_i) & \xrightarrow[\;\mathrm{can}\;]{} & X_p.
\end{array}
$$

So concluding as above, it follows that $\gamma_i \circ \delta_i = 1_{p_i}$ and that $\delta_i \circ \gamma_i = 1_{q_i}$, so we have confirmed that $\operatorname{Im}(q_i) \overset{\text{can}}{\hookrightarrow} X_p \overset{\bar{\alpha}}{\to} X$ is the image of p_i.

Combining this with our previous result, we get the following commutative diagram with an induced isomorphism γ:

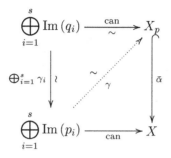

But from proposition A.1(3c) we know that the canonical morphism

$$\operatorname{Im}(p) \oplus \operatorname{Im}(p_{s+1}) \longrightarrow X$$

is an isomorphism, so we get

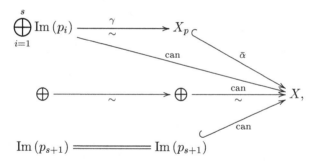

which proves that the canonical morphism

$$\bigoplus_{i=1}^{s+1} \operatorname{Im}(p_i) \longrightarrow X$$

is an isomorphism and therefore concludes the proof of the proposition. $\square$

Appendix B

Computation of $\mathcal{J}^3_{\mathbb{F}_{25}}(1,1,1)$

In this appendix, we want to compute the Jacobi sum $\mathcal{J}^3_{\mathbb{F}_{25}}(1,1,1)$. Since $\mathbb{F}_5$ does not contain the third roots of unity, we can define $\mathbb{F}_{25}$ as $\mathbb{F}_5(\zeta)$ with $\zeta^2 + \zeta + 1 = 0$. By (11.8), we have

$$
\begin{aligned}
\mathcal{J}^3_{\mathbb{F}_{25}}(1,1,1) &= (-1)^3 \sum_{\substack{(v_2,\ldots,v_3)\in(\mathbb{F}_{25}^\times)^{3-1} \\ 1+v_2+\ldots+v_n=0}} \left[\prod_{i=2}^3 \chi(v_i)^1 \right] \\
&= - \sum_{\substack{(v_2,v_3)\in(\mathbb{F}_{25}^\times)^2 \\ 1+v_2+v_3=0}} \left[\chi(v_2) \cdot \chi(v_3) \right].
\end{aligned}
\tag{B.1}
$$

In order to compute the Jacobi sum, we first have to compute the multiplicative character χ:

$$
\mathbb{F}_{25} \xrightarrow{\quad\chi\quad} \mu_3
$$

$$
x \longmapsto x^{\frac{25-1}{3}} = x^8.
$$

We do that in the following table by first computing the square of each element of $\mathbb{F}_{25}^\times$, then the fourth and finally the eighth power:

x		x^2	x^4	$x^8 = \chi(x)$
1		1	1	1
2		4	1	1
3		4	1	1
4		1	1	1
ζ	$\zeta^2 =$	$4+4\zeta$	ζ	ζ^2
$1+\zeta$	$1+2\zeta+\zeta^2 =$	ζ	$4+4\zeta$	ζ
$2+\zeta$	$4+4\zeta+\zeta^2 =$	$3+3\zeta$	4ζ	ζ^2
$3+\zeta$	$4+\zeta+\zeta^2 =$	3	4	1
$4+\zeta$	$1+3\zeta+\zeta^2 =$	2ζ	$1+\zeta$	ζ
2ζ	$4\zeta^2 =$	$1+\zeta$	ζ	ζ^2
$1+2\zeta$	$1+4\zeta+4\zeta^2 =$	2	4	1
$2+2\zeta$	$4+3\zeta+4\zeta^2 =$	4ζ	$4+4\zeta$	ζ
$3+2\zeta$	$4+2\zeta+4\zeta^2 =$	3ζ	$1+\zeta$	ζ
$4+2\zeta$	$1+\zeta+4\zeta^2 =$	$2+2\zeta$	4ζ	ζ^2
3ζ	$4\zeta^2 =$	$1+\zeta$	ζ	ζ^2
$1+3\zeta$	$1+\zeta+4\zeta^2 =$	$2+2\zeta$	4ζ	ζ^2
$2+3\zeta$	$4+2\zeta+4\zeta^2 =$	3ζ	$1+\zeta$	ζ
$3+3\zeta$	$4+3\zeta+4\zeta^2 =$	4ζ	$4+4\zeta$	ζ
$4+3\zeta$	$1+4\zeta+4\zeta^2 =$	2	4	1
4ζ	$\zeta^2 =$	$4+4\zeta$	ζ	ζ^2
$1+4\zeta$	$1+3\zeta+\zeta^2 =$	2ζ	$1+\zeta$	ζ
$2+4\zeta$	$4+\zeta+\zeta^2 =$	3	4	1
$3+4\zeta$	$4+4\zeta+\zeta^2 =$	$3+3\zeta$	4ζ	ζ^2
$4+4\zeta$	$1+2\zeta+\zeta^2 =$	ζ	$4+4\zeta$	ζ

Now we can use this table of values of χ to compute the sum from formula (B.1):

v_2	v_3	$\chi(v_2)$	$\chi(v_3)$	$\chi(v_2) \cdot \chi(v_3)$
0	4			
1	3	1	1	1
2	2	1	1	1
3	1	1	1	1
4	0			
ζ	$4 + 4\zeta$	ζ^2	ζ	1
$1 + \zeta$	$3 + 4\zeta$	ζ	ζ^2	1
$2 + \zeta$	$2 + 4\zeta$	ζ^2	1	ζ^2
$3 + \zeta$	$1 + 4\zeta$	1	ζ	ζ
$4 + \zeta$	4ζ	ζ	ζ^2	1
2ζ	$4 + 3\zeta$	ζ^2	1	ζ^2
$1 + 2\zeta$	$3 + 3\zeta$	1	ζ	ζ
$2 + 2\zeta$	$2 + 3\zeta$	ζ	ζ	ζ^2
$3 + 2\zeta$	$1 + 3\zeta$	ζ	ζ^2	1
$4 + 2\zeta$	3ζ	ζ^2	ζ^2	ζ
3ζ	$4 + 2\zeta$	ζ^2	ζ^2	ζ
$1 + 3\zeta$	$3 + 2\zeta$	ζ^2	ζ	1
$2 + 3\zeta$	$2 + 2\zeta$	ζ	ζ	ζ^2
$3 + 3\zeta$	$1 + 2\zeta$	ζ	1	ζ
$4 + 3\zeta$	2ζ	1	ζ^2	ζ^2
4ζ	$4 + \zeta$	ζ^2	ζ	1
$1 + 4\zeta$	$3 + \zeta$	ζ	1	ζ
$2 + 4\zeta$	$2 + \zeta$	1	ζ^2	ζ^2
$3 + 4\zeta$	$1 + \zeta$	ζ^2	ζ	1
$4 + 4\zeta$	ζ	ζ	ζ^2	1
$\sum$				$11 + 6\zeta + 6\zeta^2 = 5$

So it follows from formula (B.1) that $\boxed{\mathcal{J}^3_{\mathbb{F}_{25}}(1,1,1) = -5}$.

Bibliography

Siegfried Bosch. *Algebra.* Springer, Berlin, Heidelberg, New York, 1993.

Lars Brünjes. Fast-étale Überlagerungen und Faltings' Reinheitssatz in den Dimensionen eins and zwei. Diplomarbeit, Universität zu Köln, 1998.

Pierre Deligne. La conjecture de Weil. I. *Institut des Hautes Études Scientifiques. Publications Mathématiques,* (43):273–307, 1974.

P. Deligne. *Hodge Cycles, Motives and Shimura Varieties,* volume 900 of *Lecture Notes in Mathematics.* Springer, Berlin, Heidelberg, New York, 1982.

P. Deligne and N. Katz. *Groupes de Monodromie en Géométrie Algébrique (SGA 7 II),* volume 340 of *Lecture Notes in Mathematics.* Springer, Berlin, Heidelberg, New York, 1973.

E. Freitag and Reinhardt Kiehl. *Étale Cohomology and the Weil Conjecture,* volume 13 of *Ergebnisse der Mathematik und ihrer Grenzgebiete, 3. Folge.* Springer, New York, Berlin, Heidelberg, 1988.

A. Grothendieck, M. Artin, and P. Deligne. *Théorie des topos et cohomologie étale des schémas (SGA 4 III),* volume 305 of *Lecture Notes in Mathematics.* Springer, Berlin, Heidelberg, New York, 1973.

A. Grothendieck, P. Deligne, J. F. Boutot, L. Illusie, and J. L. Verdier. *Cohomologie Étale (SGA 4 1/2),* volume 569 of *Lecture Note in Mathematics.* Springer, Berlin, Heidelberg, New York, 1977.

A. Grothendieck, J.-P. Jouanolou, L. Illusie, and I. Bocur. *Cohomologie l-adique et Fonctions L (SGA 5),* volume 589 of *Lecture Note in Mathematics.* Springer, Berlin, Heidelberg, New York, 1977.

Israel M. Gelfand, Mikhail M. Kapranov, and Andrei V. Zelevinsky. *Discriminants, Resultants and Multidimensional Determinants.* Mathematics: Theory & Applications. Birkhäuser, Boston, Basel, Berlin, 1994.

A. Grothendieck. Éléments de géométrie algébrique: IV. Étude locale des schémas et des morphism de schémas, Troisième partie. *Publications mathématiques de l'I.H.É.S.,* 28(2):5–255, 1966.

A. Grothendieck. *Revêtements étales et groupe fondamental (SGA 1),* volume 224 of *Lecture Notes in Mathematics.* Springer, New York, Berlin, Heidelberg, 1971.

A. Grothendieck and J. L. Verdier. *Théorie des topos et cohomologie étale des schémas (SGA4 I)*, volume 269 of *Lecture Notes in Mathematics*. Springer, Berlin, Heidelberg, New York, 1972.

A. Grothendieck, J. L. Verdier, and B. Saint-Donat. *Théorie des topos et cohomologie étale des schémas (SGA 4 II)*, volume 270 of *Lecture Notes in Mathematics*. Springer, Berlin, Heidelberg, New York, 1973.

Fernando Q. Gouvêa and Noriko Yui. *Arithmetic of Diagonal Hypersurfaces over Finite Fields*, volume 209 of *London Mathematical Society, Lecture Note Series*. Cambridge University Press, Cambridge, 1995.

Robin Hartshorne. *Algebraic Geometry*, volume 52 of *Graduate Texts in Mathematics*. Springer, New York, Berlin, Heidelberg, 1993.

J. William Hoffman and Jorge Morales. Arithmetic of Binary Cubic Forms. *L'Enseignement Mathématique*, 46:61–94, 2000.

N. M. Katz. Review of l-adic Cohomology. In *Motives*, volume 55, Part I of *Proceedings of Symposia in Pure Mathematics*, pages 21–30, 1991.

S. L. Kleiman. The Standard Conjectures. In *Motives*, volume 55, Part I of *Proceedings of Symposia in Pure Mathematics*, pages 3–20, 1991.

Anthony W. Knapp. *Elliptic Curves*. Princeton University Press, Princeton, New Jersey, 1992.

Reinhardt Kiehl and Rainer Weissauer. *Weil Conjectures, Perverse Sheaves and l-adic Fourier Transform*, volume 42 of *Ergebnisse der Mathematik und ihrer Grenzgebiete, 3. Folge*. Springer, New York, Berlin, Heidelberg, 2001.

Serge Lang. *Algebra*. Addison-Wesley Publishing Company, Reading (Massachusetts), Menlo Park (California), New York, 1993.

J. S. Milne. *Étale Cohomology*. Princeton University Press, Princeton, New Jersey, 1980.

David Mumford. *The Red Book of Varieties and Schemes*, volume 1358 of *Lecture Notes in Mathematics*. Springer, New York, Berlin, Heidelberg, 1994.

Christopher Rupprecht. Kohomologische Invarianten für Formen höheren Grades. Diploma thesis, Universität zu Köln, 1996.

Christopher Rupprecht. *Cohomological Invariants for Higher Degree Forms*. PhD thesis, Universität Regensburg, OPUS Regensburg, http://www.opus-bayern.de/uni-regensburg/volltexte/2003/259/, July 2003.

Jean-Pierre Serre. *A Course in Arithmetic*, volume 7 of *Graduate Texts in Mathematics*. Springer, New York, Berlin, Heidelberg, 1973.

Jean-Pierre Serre. *Local Fields*, volume 67 of *Graduate Texts in Mathematics*. Springer, New York, Berlin, Heidelberg, 1979.

Jean-Pierre Serre. *Galois Cohomology*. Springer, New York, Berlin, Heidelberg, 1997.

Tetsuji Shioda. The Hodge conjecture for Fermat Varieties. *Mathematische Annalen*, 245(2):175–184, 1979.

Tetsuji Shioda. Geometry of Fermat Varieties. *Number Theory Related to Fermat's Last Theorem, Progress in Math.*, 26, 1982.

Tetsuji Shioda. What is known about the Hodge Conjecture? *Advanced Studies in Pure Mathematics, Algebraic varieties and analytic varieties, Proc. Symp., Tokyo 1981*, 1:55–68, 1983.

Tetsuji Shioda. Some Observations on Jacobi Sums. *Advanced Studies in Pure*

Mathematics 12, Galois Representations and Arithmetic Algebraic Geometry, pages 119–135, 1987.

Tetsuji Shioda. Arithmetic and geometry of Fermat curves. In *Algebraic Geometry Seminar, Singapore, 1987*, pages 95–102. World Sci. Publishing, Singapore, 1988.

Tetsuji Shioda and Toshiyuki Katsura. On Fermat Varieties. *Tôhoku Math. Journ.*, 31:97–115, 1979.

Andrzej Sładek and Adam Wesołowski. Clifford-Littlewood-Eckmann groups as orthogonal groups of forms of higher degree. *Annales Mathematicae Silesianae*, 12:93–103, 1998.

Günter Tamme. *Introduction to Étale Cohomology*. Universitext. Springer, New York, Berlin, Heidelberg, 1994.

André Weil. Numbers of Solutions of Equations in Finite Fields. *Bulletin of the American Mathematical Society*, 55(1):497–508, 1949.

André Weil. Jacobi Sums as "Größencharaktere". *Transactions of the American Mathematical Society*, 73:487–495, 1952.

Adam Wesołowski. Automorphism and similarity groups of forms determined by the characteristic polynomial. *Communications in Algebra*, 27(7):3109–3116, 1999.

Index

algebraic cycle, 79, 85, 199

base change, 6, 17, 19, 25
  conjecture, *see* Weil conjectures,
    base change
  in étale cohomology, 88
Betti number, 14
binary cubic equation, 4, 8, 135–152,
  162, 209

calibration, 80, 83
Cardano, formula of, 145
category
  fibre-, 20
  fibred-, 19–22
  filtered-, 72
  of (discrete) S-groups, 35–74, 166,
    178, 179
  of (discrete) S-sets, 34, 35, 57, 62
  of coefficient extensions, *see*
    coefficient extension
  of discrete G_k-modules, 77
  of discrete S-modules, 33
  of étale sheaves, 75
  of graded anticommutative
    L-algebras, 78
  of Grothendieck motives, *see* motif
  of k-varieties, *see* variety
  of morphisms, 21
  of non-singular equations, 135
  of pointed sets, 36–74, 94, 95, 100,
    127, 157

of S-groups, *see* S-group
  pseudo-abelian-, 9, 165, 170, 175,
    222–226
  R-linear-, 170, 175, 217–226
character, 9, 109, 165, 167–170, 173,
  175, 177–179, 187, 188, 205, 206
  cyclotomic-, 89
  Hecke-, *see* Hecke character
  induced-, 206
  multiplicative-, 186, 188, 227
  unit-, 165, 167
characteristic polynomial, 87, 88, 90,
  211, 214, 215
Chow group, 79
cocycle, 6, 37–74, 94–98, 100–102,
  107–113, 115–119, 125, 129, 159,
  160, 213
  trivial-, 38, 47, 100, 110
cocycle condition, 38–40, 43, 46, 55,
  61, 64, 100
coefficient extension, 5–7, 9, 15–33,
  36, 91–102, 156
  continuous-, 31–32, 34, 96, 97, 101,
    156
cohomologous, 38, 40, 41, 43, 46, 47,
  55, 56, 58–62, 64–66, 70, 71, 95,
  100, 110
cohomology
  étale-, 14, 75, 76, 88, 103
  Galois-, 108, 143–144
  group, 33, 96
  l-adic-, 2, 3, 5–7, 9, 10, 14, 75,